I0836571

BUFFALO

1. THE BUFFALO AT HOME

2. THE OLD TRAINS

LAND

'69

ROADMAKER

A SKELETON PATH

3. THE NEW TRAINS.

4. THE CITY

BUREAU OF ILLUSTRATION BUFFALO, N.Y.

BUFFALO LAND:

AN

AUTHENTIC ACCOUNT

OF THE

Discoveries, Adventures, and Mishaps of a Scientific and Sporting Party

IN THE WILD WEST;

WITH

GRAPHIC DESCRIPTIONS OF THE COUNTRY; THE RED MAN, SAVAGE AND CIVILIZED; HUNTING THE BUFFALO, ANTELOPE, ELK, AND WILD TURKEY; ETC., ETC.

REPLETE WITH INFORMATION, WIT, AND HUMOR.

The Appendix Comprising a Complete Guide for Sportsmen and Emigrants.

BY

W. E. WEBB,

OF TOPEKA, KANSAS.

Profusely Illustrated

FROM ACTUAL PHOTOGRAPHS, AND ORIGINAL DRAWINGS BY HENRY WORRALL.

CINCINNATI AND CHICAGO:

E HANNAFORD & COMPANY.

SAN FRANCISCO: F. DEWING & CO.

1872.

STEREOTYPED AT THE FRANKLIN TYPE FOUNDRY, CINCINNATI.

TO

The Primeval Man,

The Original Westerner, and First Buffalo Hunter,

This Work is Dedicated,

WITH PROFOUND REGARD,

BY THE AUTHOR.

BUFFALO LAND.

BY OUR TAMMANY SACHEM.

THERE'S a wonderful land far out in the West,
  Well worthy a visit, my friend;
There, Puritans thought, as the sun went to rest,
  Creation itself had an end.
'T is a wild, weird spot on the continent's face,
  A wound which is ghastly and red.
Where the savages write the deeds of their race
  In blood that they constantly shed.
The graves of the dead the fair prairies deface,
  And stamp it the kingdom of dread.

The emigrant trail is a skeleton path;
  You measure its miles by the bones;
There savages struck, in their merciless wrath,
  And now, after sunset, the moans,
When tempests are out, fill the shuddering air,
  And ghosts flit the wagons beside,
And point to the skulls lying grinning and bare
  And beg of the teamsters a ride;
Sometimes 't is a father with snow on his hair,
  Again, 'tis a youth and his bride.

What visions of horror each valley could tell,
  If Providence gave it a tongue!
How often its Eden was changed to a hell,
  In which a whole train had been flung;

How death cry and battle-shout frightened the birds,
  And prayers were as thick as the leaves,
And no one to catch the poor dying one's words
  But Death, as he gathered his sheaves:
You see the bones bleaching among the wild herds,
  In shrouds that the field spider weaves.

That era is passing—another one comes,
  The era of steam and the plow,
With clangor of commerce and factory hums,
  Where only the wigwam is now.
Like mist of the morning before the bright sun,
  The cloud from the land disappears;
The Spirit of Murder his circle has run
  And fled from the march of the years;
The click of machine drowns the click of the gun,
  And day hides the night time of tears.

PREFACE.

THE purpose of this work is to make the reader better acquainted with that wild land which he has known from childhood, as the home of the Indian and the buffalo. The Rocky Mountain chain, distorted and rugged, has been aptly called the colossal vertebræ of our continent's broad back, and from thence, as a line, the plains, weird and wonderful, stretch eastward through Colorado, and embrace the entire western half of Kansas.

Fortune, not long since, threw in my way an invitation, which I gladly accepted, to join a semi-scientific party, since somewhat known to fame through various articles in the newspaper press, in a sojourn of several months on the great plains. At a meeting held with due solemnity on the eve of starting, the Professor (to whom the reader will be introduced in the proper connection) was chosen leader of the expedition, while to my lot fell the office of editor of the future record, or rather Grand Scribe of what we were pleased to call our "Log Book" The latter now lies before me, in all its glory of shabby covers and dirty pages. Its soiled face is as honorable as that of the laborer who comes from his task in a well harvested field. Out of the sheaves gathered during our journey, I shall try and take such portions as may best supply the mental cravings of the countless thousands who hunger for the life and the lore of the far West.

I have given the mistakes as well as triumphs of our expedition, and the members of the party will readily recognize their familiar camp names. The disguise will probably be pleasant, as few like to see their failures on public parade, preferring rather to leave these in barracks, and let their successes only appear at review.

The plains have a face, a people, and a brute creation, peculiarly their own, and to these our party devoted earnest study. The expedition presented a rare opportunity of becoming acquainted with the game of the country; and, in writing the present volume, my aim has been to make it so far a text-book for amateur hunters that they may become at once conversant with the habits of the game, and the best manner of killing it. The time is not far distant, when the plains and the Rocky

Mountains will be sought by thousands annually, as a favorite field for sport and recreation.

Another and still larger class, it is hoped, will find much of interest and value in the following pages. From every state in the Union, people are constantly passing westward. We found emigrant wagons on spots from which the Indians had just removed their wigwams. Multitudes more are now on the way, with the earnest purpose of founding homes and, if possible, of finding fortunes. In order to aid this class, as well as the sportsman, I have gathered in an appendix such additional information as may be useful to both.

The scientific details of our trip will probably be published in proper form and time, by the savans interested. In regard to these, my object has been simply to chronicle such matters as made an impression upon my own mind, being content with what cream might be gathered by an amateur's skimming, while the more bulky milk should be saved in capacious scientific buckets.

Professor Cope, the well known naturalist, of the Academy of Sciences, Philadelphia, received for examination and classification the most valuable fossils we obtained, and to him I am indebted for a large amount of most interesting and valuable scientific matter, which will be found embodied in chapters twenty-third and twenty-fourth.

The illustrations of men and brutes in this work are studies from life. Whenever it was possible, we had photographs taken.

The plains, it must be said, are a tract with which Romance has had much more to do than History. Red men, brave and chivalrous, and unnatural buffalo, with the habits of lions, exist only in imagination. In these pages, my earnest endeavor, when dealing with actualities, has been to "hold the mirror up to Nature," and to describe men, manners, and things as they are in real life upon the frontiers, and beyond, to-day.

W. E. W.

TOPEKA, KANSAS, *May*, 1872.

CONTENTS.

CHAPTER I.

CHAPTER II.

CHAPTER III.

CHAPTER IV.

CHAPTER V.

CHAPTER VI.

CHAPTER VII.

CHAPTER VIII.

CHAPTER IX.

CHAPTER X.

CHAPTER XI.

CHAPTER XII.

CHAPTER XIII.

CHAPTER XIV.

CHAPTER XV.

CHAPTER XVI.

CHAPTER XVII.

CHAPTER XVIII.

CHAPTER XIX.

CHAPTER XX.

CHAPTER XXI.

CHAPTER XXII.

CHAPTER XXIII.

CHAPTER XXIV.

CHAPTER XXV.

CHAPTER XXVI.

CHAPTER XXVII.

CHAPTER XXVIII.

CHAPTER XXIX.

CONTENTS OF APPENDIX.

LIST OF ILLUSTRATIONS.

From Original Drawings by Henry Worrall, and Actual Photographs.
The Engraving by the Bureau of Illustration, Buffalo, N. Y.

BUFFALO LAND.

CHAPTER I.

THE OBJECT OF OUR EXPEDITION—A GLIMPSE OF ALASKA THROUGH CAPTAIN WALRUS' GLASS—WE ARE TEMPTED BY OUR RECENT PURCHASE—ALASKAN GAME OF "OLD SLEDGE"—THE EARLY STRUGGLES OF KANSAS—THE SMOKY HILL TRAIL—INDIAN HIGH ART—THE "BORDER-RUFFIAN," PAST AND PRESENT—TOPEKA—HOW IT RECEIVED ITS NAME—WAUKARUSA AND ITS LEGEND.

THE great plains—the region of country in which our expedition sojourned for so many months—is wilder, and by far more interesting, than those solitudes over which the Egyptian Sphynx looks out. The latter are barren and desolate, while the former teem with their savage races and scarcely more savage beasts. The very soil which these tread is written all over with a history of the past, even its surface giving to science wonderful and countless fossils of those ages when the world was young and man not yet born.

At first, it was rather unsettled which way the steps of our party would turn; between unexplored territory and that newly acquired, there were several fields open which promised much of interest. Originally, our company numbered a dozen; but Alaska tempted a portion of our savans, and to the fishy and frigid maiden they yielded, drawn by a strange predilection for train-oil and seal meat toward the land of furs. For the remainder of our party, however, life under the Alaskan's tent-pole had no charms. Our decision may have been influenced somewhat by the seafaring man with whom our friends were to sail. The real name of this son of Neptune was Samuels, but our party called him, as it savored more of salt water, Captain Walrus, of the bark Harpoon. This worthy, according to his own statement, had been born on a whaler, weaned among the Esquimeaux, and, moreover, had frozen off eight toes "trying to winter it at our recent purchase." He evidently disliked to have scientific men aboard, intent on studying eclipses and seals. "A heathenish and strange people are the Alaskans," Walrus was wont to say. "What is not Indian is Russian, and a compound of the latter and aboriginal is a mixture most villainous. One portion of the partnership anatomy takes to brandy, while the other absorbs train-oil, and so a half-breed Alaskan heathen is always prepared for spontaneous combustion, and if rubbed the wrong way, flames up instantly. He is always hot for murder, and if you throw cold water on his designs, his oily nature sheds it."

And many a yarn did the captain spin concerning their strange customs. Sealing a marriage contract consisted in the warrior leaving a fat seal at the hole of the hut, where his intended crawled in to her home privileges of smoke and fish. Their favorite game was "old sledge," played with prisoners to shorten their captivity.

All this, and much more, probably equally true, we had picked up of Alaskan history, and at one time our chests had been packed for a voyage on the Harpoon; but at the final council the west carried it

ALASKAN LOVERS—SEALING THE CONTRACT.

ALASKAN GAME OF OLD SLEDGE.

against the north, and our steps were directed toward the setting sun, instead of the polar star.

The expedition afforded unexcelled facilities for seeing Buffalo Land. It was composed of good material, and pursued its chosen path successfully, though under difficulties which would have turned back a less determined party.

None of our company, I trust, will consider it an unwarrantable license which recounts to others the personal peculiarities and mistakes about which we joked so freely while in camp. It was generally understood, before we parted, that the adventures should be common stock for our children and children's children. Why should not the great public share in it also?

Let the reader place before him a checker-board, and allow it to represent Kansas, whose shape and outline it much resembles; the half nearest him will stand for the eastern or settled portion of the State, of which the other half is embraced in Buffalo Land proper. It is with the latter that we have first to do, as with it we first became acquainted.

Our party entered the State at Kansas City, and took the cars for Topeka, its capital. During our morning ride through the valley of the Kaw, memory went backward to the years when "Bleeding Kansas" was the signal-cry of emancipation. When gray old Time, a decade and a half ago, was writing the history of those bright children of Freedom, the united sisterhood, a virgin arm reached over his shoulder, and a fair young hand, stained with its own life-blood, wrote on the page toward which all the world was gazing, "I am Kansas, latest-born of America. I would be free, yet they would make me a slave. Save me, my sisters!" The great heart of our nation was sorely distressed. Conscience pointed to one path—Policy, that rank hypocrite, to another.

And so it was that the young queen, with her grand domain in the West, struggled forward to lay her fealty at the feet of our great mother, Liberty. She made a body-guard of her own sons, and their number was quickly swelled by brave hearts from the north, east, and west. The new territory, begging admission as a State, became a battle-ground. Slavery had reached forth its hand to grasp the new State and fresh soil, but the mutilated member was drawn back with wounds which soon reached, corrupted and destroyed the body. In this land of the Far West a nation of young giants had been suddenly developed, and Kansas was forever won for freedom.

But there was yet another enemy and another danger. Westward, toward Colorado, the savage's tomahawk and knife glittered, and struck among the affrighted settlements. *Ad Astra per Aspera*, "to the stars through difficulties," the State exclaims on the seal, and to the stars, through blood, its course has been.

Those old pages of history are too bloody to be brought to light in the bright present, and we purpose turning them only enough to gather what will be now of practical use. Kansas suffered cruelly, and brooded over her wrongs, but she has long since struck hands with her bitterer foe. Most of the "Border

Ruffians" ripened on gallows trees, or fell by the sword, years ago. A few, however, are yet spared, to cheer their old age by riding around in desolate woods at midnight, wrapped in damp nightgowns, and masked in grinning death-heads. Although the mists of shadow-land are chilling their hearts, yet those organs, at the cry of blood, beat quick again, like regimental drums, for action.

The Kaw or the Kansas River, the valley of which we were traversing, is the principal stream of the State—in length to the mouth of the Republican one hundred and fifty miles, and above that, under the name of Smoky Hill, three hundred miles more.

The "Smoky Hill trail" is a familiar name in many an American home. It was the great California path, and many a time the demons of the plain gloated over fair hair, yet fresh from a mother's touch and blessing. And many a faint and thirsty traveler has flung himself with a burst of gratitude on the sandy bed of the desolate river, and thanked the Great Giver of all good for the concealed life found under the sand, and with the strength thus sucked from the bosom of our much-abused mother, he has pushed onward until at length the grand mountains and great parks of Colorado burst upon his delighted vision.

About noon we arrived at Topeka, the capital, well situated on the south bank of the river, having a comfortable, well-to-do air, which suggests the quiet satisfaction of an honest burgher after a morning of toil. The slavery billow of agitation rolled even thus far from beyond the border of the state. Armed men rode over the beautiful prairies, some east, some west—one band to transplant slavery from the tainted soil of Missouri, another to pluck it up.

A small party of Free State men settled upon this beautiful prairie. South flowed the Waukarusa, south and east the Shunganunga, and west and north the Kaw or Kansas. Here thrived a bulbous root, much loved by the red man, and here lazy Pottawatomies gathered in the fall to dig it. In size and somewhat in shape, it resembled a goose egg, and had a hard, reddish brown shell, and an interior like damaged dough. The Indian gourmands ate it greedily and called it "Topeka." From the two or three families of refugee Free State men the town grew up, and from the Indian root it took its name. Its christening took place in the first cabin erected, and it is reported that a now prominent banker of the town stood sponsor, with his back against the door, refusing any egress until the name of his choice was accepted. It is even affirmed that one opposing city founder was pulled back by his coat-tail from an attempted escape up the wide chimney.

The old Indian love of commemorating events by significant names is well illustrated in Kansas. One example may be given here. Waukarusa once opposed its swollen tide to an exploring band of red men. Now, from time beyond ken, the noble savage has been illustrious for the ingenuity with which he lays all disagreeable duties upon the shoulders of the patient squaw. He may ride to their death, in free wild sport, the bison multitudes; but their skins

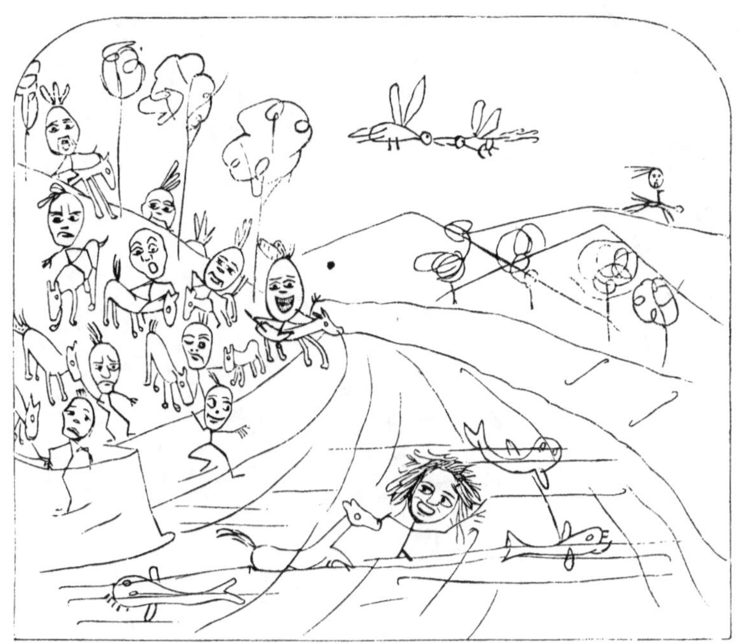

"WAUKARUSA."

"TOASTS HIS MOCCASINED FEET BY THE FIRE."

must be converted into marketable robes, and the flesh into jerked meat, by the ugly and over-worked partner of his bosom. While she pins the raw hide to earth, and bends patiently over, fleshing it with horn hatchet for weary hours, the stronger vessel, his abdominal recesses wadded with buffalo meat, toasts his moccasined feet by the fire, fills his lungs with smoke from villainous killikinick, and muses soothingly of white scalps and happy hunting grounds.

Ox-like maiden, happy "big injun!" you both belong to an age and a history well nigh past, and let us rejoice that it is so.

But to return to the band long since gathered into aboriginal dust whom we left pausing on the banks of the Waukarusa. "Deep water, bad bottom!" grunted the braves, and, nothing doubting it, one loving warrior pushed his wife and her pony over the bank to test the matter. From the middle of the tide the squaw called back, "Waukarusa" (thigh deep), and soon had gained the opposite bank in safety. Then and there the creek received its name, "Waukarusa."

We procured a remarkable sketch, in the well known Indian style of high art, commemorative of this event. It has always struck us that the savage order of drawing resembles very much that of the ancient Egyptian—except in the matter of drawing at sight, with bow or rifle, on the white man.

CHAPTER II.

A CHAPTER OF INTRODUCTIONS—PROFESSOR PALEOZOIC—TAMMANY SACHEM—DOCTOR PYTHAGORAS—GENUINE MUGGS—COLON AND SEMI-COLON—SHAMUS DOBEEN—TENACIOUS GRIPE—BUGS AND PHILOSOPHY—HOW GRIPE BECAME A REPUBLICAN.

WHEN permission was given me to draw upon the journal of our trip for such material as I might desire, it was stipulated that the camp-names should be adhered to. A company on the plains is no respecter of persons, and titles which might have caused offense before starting were received in good part, and worn gracefully thenceforward.

Our leader, Professor Paleozoic, ordinarily existed in a sort of transition state between the primary and tertiary formations. He could tell cheese from chalk under the microscope, and show that one was full of the fossil and the other of the living evidences of animal life. A worthy man, vastly more troubled with rocks on the brain than "rocks" in the pocket.

Learning had once come near making him mad, but from this sad fate he was happily saved by a somewhat Pickwickian blunder. While in Kansas, some years since, he penetrated a remote portion of the wilderness, where, as he was happy in believing, none but the native savage, or, possibly, the primeval man, could ever have tarried long enough to leave any sign behind. Imagine his astonishment and delight, therefore, when from the tangled grass he drew an upright stone, with lines chiseled on three sides and on the fourth a rude figure resembling more than any thing else one of those odd fictions which geologists call restored specimens. On a ledge near were huge depressions like foot-prints. They were foot-prints of birds, no doubt, and quite as perfect as those found in more favored localities, and from which whole skeletons had been constructed by learned men.

Both specimens were forwarded to, and at the expense of, noted savans of the East. Our professor called the pillar from the tangled grass an altar raised by early races to the winds. The short lines, he suggested, designated the different points of the compass, while the rude figure was intended for Boreas. Our scientists toward the rising sun met the boxes at the depot, paid charges, and careful draymen bore them to the expectant museum.

One hour after, seven wise men might have been seen wending their way sorrowfully homeward, with hands crossed meditatively under their coat-tails, and pocket vacuums where lately were modern coins. Government clearly had a case against our professor. Science decided that he had removed a stone telling in surveyors' signs just what section and township it was on. The figure which he had imagined a heathen idea of Boreas was the fancy of some surveyor's idle moment—a shocking sketch of an impossible buffalo. Whether the bird-tracks had a common origin, or were hewn by the hatchets of the red man, is a point still under discussion.

A worthy man, as before remarked, was the professor, full of knowledge, genial in camp, and, having rubbed his eye-tooth on a section stone, geological authority of the highest order. When the professor said a particular rock belonged to the cretaceous formation, one might safely conclude that no modern influences had been at work either on that rock or in that vicinity. That question was settled.

Next came Tammany Sachem, our heavy weight and our mystery. Before joining our party, he had been a New York alderman, noted for prowess in annual aldermanic clam-bakes at Coney Island. He was wont to exhibit a medal, the prize of such a tournament, on which several immense clams were racing to the griddle, for the honor of being devoured by the city fathers.

A green-ribbed hunting coat traversed his rotundity, which had the generous swell of a puncheon. His face was reddish, and his nose like a beacon-light against a sunset sky. When you thought him awake, he was half asleep; when you thought him asleep, he was wide awake. A look of extreme happiness always beamed on his face when misfortunes impended. Per contra, successes made him suspicious and morose. New York aldermen have always been a puzzle to the nation at large. Perhaps our friend's facial contradictions, put on originally as one of the tricks of the trade, had become chronic from long usage. We have since learned that the sachems of Tammany laugh the loudest and joke the most freely when under affliction.

When I was appointed editor, the Sachem volun-

THE PROFESSOR—A REMARKABLE STONE.

TAMMANY SACHEM—PROSPECTIVE AND RETROSPECTIVE.

teered as local reporter. Many of the items he gathered are entered in our log-book in rhyme, and to these pages some of them are transferred verbatim. In wooing the muses, our alderman certainly acted out of character. The ideal poet is thin instead of obese, and he is a reckless innovator who lays claim to any measure of the divine afflatus without possessing either a pale face, thin form, or a garret.

As to what drove a New York alderman to the society of buffaloes, we had but one explanation, and that was Sachem's own. We knew that he disliked women in every form, Sorosis and Anti-Sorosis, bitter and sweet alike. According to his statement, made to us in good faith, and which I chronicle in the same, Cupid had once essayed to drive a dart into Sachem's heart, but, in doing so, the barb also struck and wounded his liver. As his love increased, his health failed. His liver became affected in the same ratio as his heart. This was touching our alderman in a tender spot. Imagine a New York city father without digestion; what a subject of scorn he would become to his constituency! Our alderman fled from Cupid, clams, and his beloved Gotham, and sought health and buffalo on the plains of Kansas. As he remarked to us pathetically: "A good liver makes a good husband. Indigestion frightens connubial bliss out of the window. Pills, my boy, pills is the quietus of love. If you wish Cupid to leave, give him a dose of 'em. The liver, instead of the heart, is at the bottom of half the suicides."

Doctor Pythagoras in years was fifty, and in stature

short. His favorite theory was "development," and this he carried to depths which would have astonished Darwin himself. How humble he used to make us feel by digging at the roots of the family tree until its uttermost fiber lay between an oyster and a sponge! (Rumor charged him with waiting so long for diseases to develop, that his patients developed into spirits.) While he indorsed Darwin, however, he also admired Pythagoras. The latter's doctrine of metempsychosis he Darwinized. In their transmigration from one body to another, souls developed, taking a higher order of being with each change, until finally fitted to enter the land of spirits. The soul of a jack-of-all-trades was one which developed slowly, and picked up a new craft with each new body. Like Pythagoras, he remembered several previous bodies which his soul had animated, among others that of the original Rarey, who existed in Egpyt some centuries before the modern usurper was born. If souls proved entirely unworthy during the probationary or human period, they were cast back into the brute creation to try it over again. To this class belonged prize-fighters, Congressmen, and the like. With them the past was a blank—an unsuccessful problem washed from the slate. The doctor had a hobby that a vicious horse was only a vicious man entered into a lower order of being. To demonstrate this he had traveled, and still persisted in traveling, on eccentric horses, for the purpose of reasoning with them. But his Egyptian lore had been lost in transmission, and his falls, kicks, and bites became as many as the moons which had passed over his head.

COLON AND SEMI-COLON.

DAVID PYTHAGORAS, M. D.

Genuine Muggs was an Englishman. The antipodes of Tammany Sachem, who would not believe any thing, Muggs swallowed every thing. He had already absorbed so much in this way that he knew all about the United States before visiting it. Given half a chance, he would undoubtedly have told the savage more about the latter's habits than the aborigine himself knew. It was positively impossible for him to learn any thing. His round British body was so full of indisputable facts that another one would have burst it. In the Presidential alphabet, from Alpha Washington to Omega Grant, he knew all of our rulers' tricks and trades, and understood better the crooked ways of the White House than our own talented Jenkins.

British phlegm incased his soul, and British leather his feet. From heel to crown he was completely a Briton. His mutton-chop whiskers came just so far, and the h's dropped in and out of his utterings in a perfectly natural way. In the Briton's alphabet, Sachem used to remark, the *I* is so big that it is no wonder the *H* is often crowded out.

Muggs was a fair representative of the average Englishman who has traveled somewhat. The eyeteeth of these persons are generally cut with a slash, and they are forever after sore-mouthed. For a maiden effort they never suck knowledge gently in, but attempt a gulp which strangles. The consequence of this hasty acquiring is a bloated condition. The partly-traveled Briton seems, at first acquaintance, full and swollen with knowledge; but should the student of learning apply the prick, the result obtained will generally prove to be—gas.

Over our great country, some of the family of Muggs meet one at every turn. Often they scurry along solitarily, but occasionally in groups. In the former case they are unsocial to every body—in the latter to every body except their own party. The bliss which comes from ignorance must be of a thoroughly enjoyable nature, for the Muggses certainly do enjoy themselves. They will pass through a country, remaining completely uncommunicative and self-wrapped, and know less of it after six months' traveling than an American in two. The professor says he has met them in the lonely parks of the Rocky Mountains and in the fishing and hunting solitudes of the Canadas. If they have been an unusually long time without seeing a human being, they may possibly catch at an eye-glass and fling themselves abruptly into a few remarks. But it is in a tone which says, plainer than words, "No use in your going any further, man; I have absorbed all the beauties and knowledge of this locality."

It is a rare treat to see a coach delivered of Muggs at a country inn. "Hi, porter, look hout for my luggage, you know. Tell the publican some chops, rare, and lively now, and a mug of hale, and, if I can 'ave it, a room to myself." If the latter request is granted, and you are inquisitive enough to take a peep, you may see Muggs sturdily surveying himself in the glass, and giving certain satisfied pats to his cravat and waistcoat, as if to satisfy them that they covered a Briton. Could the mirror which reflects

ONE OF THE MUGGSES.

his face also reflect his thoughts, they would read about as follows: "Muggs, you are a Briton, and this hotel must be made aware of the fact. Whatever you do, be guilty of no un-English act while in this outlandish land. Your skin is now full of knowledge, and let not other travelers, like so many mosquitoes, suck it from you. Your forefathers blessed their eyes and dropped their h's, and so must you." And perhaps by this time, if the chops have arrived, he dines in seclusion and, by so doing, loses a fund of information which his fellow-travelers have obtained by common exchange.

Again on the way, Muggs nestles in a corner of the coach and acts strictly on the defensive, indignantly withdrawing his square-toed, thick-soled English shoes, should neighboring feet attempt to hobnob with them. On a trip through Buffalo Land, however, it is difficult for one of her Britannic Majesty's subjects to maintain the national dignity. But this fact Genuine Muggs—our Muggs—evidently did not know. Had he known it, he would never have gone with us in the world. 49

Another of our party rejoiced in the appellation of "Colon." He obtained this title because his eccentric specialities of character several times came very near putting if not a full stop, at least the next thing to it, upon the particular page of history which our party was making. Longitudinally, Mr. Colon was all of five feet eleven; in circumference, perhaps a score or so of inches. He possessed a fair share of oddities, and what is better an equally fair one of dollars. The hemispheres of his philanthropic brain seemed equally pre-empted by philosophy and bugs. Engaging in some immense work for the amelioration of mankind, he would pursue it with ardor, dwell upon it with unction, and then suddenly leave it, half finished, to capture a rare spider. Philosophy and Entomology had constant combat for Colon, and victory tarried with neither long enough for the seat of war to be cultivated and blossom with any luxuriance. At the time he joined our party one of his grandest charitable projects had lately died in a very early period of infancy, entirely supplanted in his affections for the time being by the prospect of a chase after Brazilian insects. During our journey it was no uncommon thing for us to see his thin form all covered with bugs and reptiles, which had crawled out of the collecting boxes carried in his pockets. If this meets our friend's eye, let him bear no malice, but reflect, in the language of his own invariable answer to our remonstrances, "It can't be helped." Should the public parade of his faults be disagreeable, he can suffer no more from them now than we did in the past, and may perhaps call them into closer quarters for the future.

Mr. Colon's son, of two years less than a score, we dubbed Semi-colon, as being a smaller edition, or to be exact, precisely one-half of what the senior Colon was. So perfect was the concord of the two that the junior had fallen into a chronic and to us amusing habit of answering "Ditto" to the senior's expressions of opinion. Divide the father's conversation by two, add an assent to every thing, and the result, socially considered, would be the son. It may readily be seen, therefore, why the professor for short should call him, as he nearly always did, "Semi."

Shamus Dobeen, our cook and body-servant, according to his own account, was the child of an impoverished but noble Irish family. Indeed, we doubt if any Irishman was ever promoted from shovel laborer to body-servant without suddenly remembering that he was "descinded" from a line of kings. At the time Shamus was added to the population of Ireland, the patrimonial estate had dwindled down to a peat bog. As this soon "petered out," Shamus went from the exhausted moor into the cold world. He had been by turns expelled patriot, dirt disturber on new railroads, gunner on a Confederate cruiser, and high private in a Union regiment. The position of gunner he lost by touching off a piece before the muzzle had been run out, in consequence of which part of the vessel's side went off suddenly with the gun. Captured, he readily became a Union soldier, and could, without doubt, have transformed himself into a Cheyenne, or a Patagonian, had occasion for either ever required.

While in Topeka, our party made the acquaintance of Tenacious Gripe, a well-known Kansas politician, and who attached himself to us for the trip. Every person in the State knew him, had known him in territorial times, and would know him until either the State or he ceased to be.

Flung headlong from somewhere into Kansas during the "border ruffian" period, he would probably have passed as rapidly out of it had he been allowed to do so peaceably. But as the slavery party endeavored to push him, he concluded to stick. At that particular time, he was a moderate Democrat or conservative Republican, and consequently had no particular principles. But the slavery party supposed he had, and to them accordingly he became an object of suspicion. They assumed the aggressive, and he at once resolved into a staunch Republican. Had the latter first struck him, he would have been as staunch a Democrat. And Gripe has never known how near he came to being the latter. The Republicans had just decided to order him out of the state as a border ruffian spy, when the Democrats took action and did so for his not being one. Those were troublous times. He went to the front at once in the antislavery ranks, and has stayed there ever since. Sore-headed men are apt to become famous. There were those in our late war who were kicked by adversity into the very arms of Fame.

Our friend had been in both the upper and lower houses of the State Legislature, and had rolled Congressional logs, moreover, until he was hardly happy without having his hands on one.

SHAMUS DOBEEN—HIS CARD.

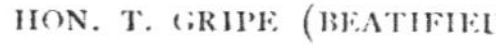
HON. T. GRIPE (BEATIFIED).

CHAPTER III.

THE TOPEKA AUCTIONEER—MUGGS GETS A BARGAIN—CYNOCEPHALUS—INDIAN SUMMER IN KANSAS—HUNTING PRAIRIE CHICKENS—OUR FIRST DAY'S SPORT.

WE had three or four days to spend in Topeka, as it was there that we were to purchase our outfit for the buffalo region. With the latter purpose in view, we were wandering along Kansas Avenue the next morning, when a horseman came furiously down the street, shouting, at the top of his lungs, "Sell um as he wars har!" Semi hastily retreated behind Mr. Colon, thinking it might be a Jayhawker, while the professor adjusted his glasses.

Muggs said the individual reminded him of the famous charge at Balaklava. Muggs had never seen Balaklava, but other Englishmen had, which answered the same purpose.

The equestrian proved to be a well-known auctioneer of Topeka, who may be discovered at almost any time tearing through the streets on some spavined or bow-legged old cob, auctioneering it off as he goes. His favorite expression is, "I'll sell um as he wars har." What particular selling charm lies concealed in this announcement even Gripe could not tell. Sachem thought that possibly he had been brought up at some exposed frontier post, where, on account of Indian prejudices, wearing hair is a rare luxury.

To say there that a man was still able to comb his own scalp-lock denoted an extraordinary state of physical perfection. Expressions of praise for humans are often applied to horses, and so, perhaps, the one in question. "I have heard," quoth our alderman, in support of this assertion, "Fitz say of a belle, at a charity ball, what a 'bootiful çweature; and I have heard him, the day after, in his stable, say the same thing of his horse."

That horse-auction was a sight worth seeing. The crowd collected most thickly on the corner of Kansas Avenue and Sixth Street, and before it the cob came to a stand. And it was a stand—as stiff and painful as that of a retired veteran put on dress parade. The limbs would have had full duty to perform in supporting the carcass alone, which had evidently been in light marching order for years past. The additional weight of the auctioneer must certainly have proved altogether too much, had not the horse heard, for the first time, of the wonderful qualities with which he was still endowed.

Seeing a whole corner, with gaping mouths, swallowing the statement that he was only six years old, reduced by hard work, and could, after three months grass, pull a ton of coal, he would have been a thankless horse indeed, which could not strain a point, or all his points, for such a rider.

And so, when the spurs suddenly rattled against his ribs, the old skin full of bones gave a snort of pain, which the auctioneer called "Sperit, gentle*men*!" and away up the broad avenue he rolled, at a speed which threatened to break the rider's neck, and his

"SPERIT, GENTLEMEN!"

own legs as well. His tail having been cut short in youth, and retrimmed in old age, the outfit made but a sorry figure going up the street. The Professor said it suggested the idea of some fossil vertabra, with a paint brush attached to its end, running away with a geological student.

After the return and cries for more bids, Muggs must have winked at the auctioneer—possibly, to slyly telegraph him the fact that in "Hengland" they were up to such games. At least the auctioneer so declared, and advancing the price one dollar in accordance therewith, finally knocked the brute down to him. Then the British wrath bubbled and boiled. The auctioneer was inexorable. Muggs *had* winked, and that was an advanced bid, according to commercial custom the land over. Articles were often sold simply by the vibration of an eyelash, and not a word uttered.

The Professor remarked that in law winks would doubtless be accepted as evidence. It was a recognized principle of the statutes that he who winked at a matter acquiesced in it, and indeed such signals were often more expressive than words. Sachem sustained this point, and added further that he had known many a man's head broken on account of an injudicious wink.

The crowd, with almost unanimous voice, pronounced the auctioneer right and Muggs wrong.

"Me take the brute!" exclaimed the indignant Briton; "why he can 'ardly stand up long enough to be knocked down. Except in France, he could be put to no earthly use whatever. 'Is knees knock together in an ague quartette, and 'is tail—look at it! It's hincapable of knocking a fly off; looks more like flying off hitself!" Muggs further declared the sale was an attempt on the owner's part to evade the health officer, who would have been around, in a couple of days, to have the carcass removed.

The auctioneer waxed belligerent, the crowd noisy, and Muggs, like a true Englishman, secured peace at the price of British gold. The horse was on his hands, having barely escaped being on the town, and an enthusiastic crowd of urchins escorted the purchase to a livery stable. Muggs christened the animal Cynocephalus, and soon afterward sold him to Mr. Colon, who was of an economical turn, for the use of his son Semi.

"I have heard," said the thoughtful father, "that the buffalo grass of the plains is very nourishing. All that the poor steed needs is care and fat pastures. Semi can give him the former, and over the latter our future journey lies. I have also learned that what is especially needed in a hunting horse is steadiness, and this quality the animal certainly possesses."

From some months' acquaintance with the purchase, we can say that Cynocephalus was steady to a remarkable degree. We are firmly persuaded that a heavy battery might have fired a salute over his back without moving him, unless, possibly, the concussion knocked him down.

Our first hunting morning, the second day preceding our hegira westward, came to us with a clear sky, the sun shedding a mellow warmth, and the air full of those exhilarating qualities which our lungs afterward drank in so freely on the plains. Indian summer, delightful anywhere, is especially so in Kansas.

From the advance guard of the winter king not a single chilling zephyr steals forward among the tarrying ones of summer. Soothing and gentle as when laden with spicy fragrance south, they here shower the whole land with sunbeams. Earth no longer seems a heavy, inert mass, but floats in that smoky, fleecy atmosphere with which artists delight so much to wrap their angels. It is as if the warmer, lighter clouds of sunny weather were nestling close to earth, frightened from the skies, like a flock of white swans, at the October howls of winter. But I never could agree with those writers who call this season dreamy. If such it be, it is surely a dream of motion. All nature appears quickened. The inhabitants of the air have commenced their southern pilgrimage, and the oldest and leading ganders may be heard croaking, day-time and night-time, to their wedge-shaped flocks their narrative of summer experiences at the Arctic circle, and their commands for the present journey.

Sachem, I find, has recorded as a discovery in natural history that geese form their flocks in wedge shape that they may easier "make a split" for the south when Nature, with her north pole, stirs up their feeding and breeding-grounds in November gales, and changes their fields of operation into fields of ice. Sachem was sadly addicted to slang phrases.

All game, I may remark, is wilder at this season of the year than earlier. If the earth is dreaming, its wild inhabitants certainly are not. Men, too, have thrown off the summer lethargy, and shave their neighbors as closely as ever. If any one thinks it a dreamy season of the year, let him test the matter practically by being a day or two behindhand with a payment.

In reply to a question, the professor told us that the smoky condition of the atmosphere was probably caused by the exhalation of phosphorus from decaying vegetation. Sachem remarked that out of twenty different objects which he had submitted for examination, and as many questions that he had asked, nine-tenths of the results contained phosphorus in some shape. It was becoming monotonous and dangerous.

While the party thus mused and speculated, we had come out into the open country, south-west of town, and were now approaching Webster's Mound, a cone-shaped hill from which we afterward obtained some excellent views. For the trip we had been supplied with two dogs, one a setter, belonging to the private secretary of the Governor, and the other a pointer, the property of a real estate dealer. The former was an ancient and venerable animal. The rheumatism was seized of his backbone and held high revel upon the juices which should have lubricated the joints. Even his tail wagged with a jerk, inclining the body to whichever side it had last swung. He was so full of rheumatism that whenever he scented a chicken the pain evoked by the excitement caused him to howl with anguish. The pointer, per contra, was hale and swift, but had lost one eye; and a shot from the same charge which destroyed that organ, rattled also on his left ear-drum, and that membrane no longer responded to the shouts of the hunter. On one side he could see, and not hear—on the other, hear, but not see. Nevertheless, with gestures for the left view, and shouts on the right, fair work might still be obtained. Both dogs rejoiced in the uncommon name of Rover, and both possessed that most excellent of all points in such animals, a steady point.

If any of my readers are fond of field-sports, and have not yet shot prairie-chickens over a dog, let them take their guns and hie to the West, and taste for themselves of this rare sport. With the wide prairie around him, keeping the bird in full view during its passage through the air, one can choose his distance for firing and witness the full effect of his shot. I think the brief instant when the flight of the bird is checked and it drops head-foremost to earth, is the sweetest moment of all to the hunter.

CHAPTER IV.

CHICKEN-SHOOTING CONTINUED—A SCIENTIFIC PARTY TAKE THE BIRDS ON THE WING—EVILS OF FAST FIRING—AN OLD-FASHIONED "SLOW SHOT"—THE HABITS OF THE PRAIRIE-CHICKEN—ITS PROSPECTIVE EXTINCTION—MODE OF HUNTING IT—THE GOPHER SCALP LAW.

WE had left the road and were now driving over the fine prairie skirting Webster's Mound, the grass being about a foot high and affording excellent cover. Taking advantage of its being matted so closely from the early frosts, the old cocks hid under the thick tufts and called for close work on the part of our dogs.

Back and forth across our path these intelligent animals ranged, the one fifty yards or so to our right, the other as many to our left, crossing and re-crossing, with open mouths drinking in eagerly the tainted breeze. This latter was in our favor, and both dogs suddenly joined company and worked up into it, with outstretched noses pointing to game that was evidently close ahead.

The pointer crawled cautiously, like a tiger, his spotted belly sweeping the earth, and his tail, which had been lashing rapidly an instant before, gradually stiffening. He would open his mouth suddenly, drink in a quick, deep draught of air, and, closing the jaws again, hold it until obliged to take another respiration. He seemed as loath to let the scent of the chicken pass from his nostrils as a hungry newsboy is to quit the savory precincts of Delmonico's kitchen window. The setter's old bones appeared to renew their youth under the excitement, and he was as active as a retired war-horse suddenly plunged into battle.

Both dogs came simultaneously to a point—tails curved up and rigid, each body motionless as if cut in marble and one forepaw lifted. No wonder so many men are wild with a passion for hunting. Kind Providence smiles upon the legitimate sport from conception to close, and gives us a *posé* to start with fascinating to any lover of the beautiful, whether hunter or not. But one must not pause to moralize while dogs are on the point, or he will have more philosophy than chickens.

All the party had got safely to ground and were behind the dogs, with guns ready and eyes eagerly fastened on the thick grass which concealed its treasure as completely as if it had been a thousand miles below its roots, or on the opposite side of this mundane sphere in China. Not a thing was visible within fifty yards of our noses save two dogs standing motionless, with stiffened tails and eyes fixed on, and nozzles pointed toward, a spot in the sea of brown, withered grass, not ten feet away.

The Professor took out his lens, Mr. Colon let down the hammers of his gun and cocked them again, to be sure all was right, while Sachem wore a puzzled expression as if undecided whether the attitude of the dogs indicated any thing particular or not. The grass nodded and rustled in the light wind, but not a blade moved to indicate the presence of any living thing beneath it, while the dogs remained as if petrified.

The Professor said it was very remarkable, and wondered what had better be done next. Mr. Colon thought that the dogs were tired, and we might as well get into the wagon. Another suggested at random that we should set the dogs on, and Muggs, who had probably heard the expression somewhere, cried, "Hi, boys, on bloods!" At the words the dogs made a few quick steps forward, and on the instant the grass seemed alive with feathered forms, popping into air like bobs in shuttlecock. Such a fluttering and flying I have never seen since, when a boy, I ventured into a dove cote, and was knocked over by the rush of the alarmed inmates. From under our very feet, almost brushing our faces, the beautiful pinnated grouse of the prairies left their cover, and us also.

Every gun had gone off on the instant, and we doubt if one was raised an inch higher than it happened to be when the covey started. The Professor afterward extracted some stray shot from the legs of his boots, and the setter, which was next to Muggs, gave a cry of pain for which there was evidently other cause than rheumatism, as was demonstrated by his retirement to the rear, from which he refused to budge until we all got into the wagon, and to which he invariably retreated whenever we got out.

From the midst of the birds which were soaring away, one was seen to rise suddenly a few feet above

OUR FIRST BIRD-SHOOTING.

his comrades, and then fall straight as a plummet, and head first, to earth. It had caught some stray shot from the bombardment—Muggs claimed from his gun, but this statement the setter, could he have spoken, would certainly have disputed.

Semi-Colon brought in the game, which proved to be a fine male, with whiskers and full plumage, which must have made sad havoc among the hearts of the hens, when the old fellow was on annual dress parade in the spring. At that season of the year the cocks seek some knoll of the prairie, where the grass has been burnt or cut off, and strut up and down with ruffled feathers, uttering meanwhile a booming sound, which can be heard in a clear morning for miles. The flabby pink skin that at other seasons hangs in loose folds on his neck is then distended like a bagpipe, and he is a very different bird from the same individual in his Quaker gray and respectable summer and fall habits.

Ensconced again in the wagon, our party moved forward, the dogs, as before, examining the prairie. The professor was comparing the birds of the present and the past ages, when Muggs suddenly blasted his eyes and declared the beasts were at it again. And so they were, the setter making a good stand at some game in the grass, and the other dog, a short distance off, pointing his companion. During the remainder of the day we found many large flocks of birds, and fired away until two or three swelled noses testified how dirty our guns were.

"Fast shooting," said the professor, as we were on our way home, "is as bad as that too slow. Although I am no sportsman from practice, I love and have studied the principles of it. In my father's day the rule was, when a bird rose, for a hunter to take out his snuff-box, take snuff, replace the box, aim, and fire. You may find the advice yet in some works. The shot then has distance in which to spread. With close shooting they are all together, and you might as well fire a bullet. When you have given the bird time, act quickly. The first sight is the best. Again, the first moment of flight, with most birds, is very irregular, as it is upward, instead of from you."

Dobeen begged leave to inform our "honors" that in Ireland, after a bird rose, the rule was, instead of taking snuff, to take off the boots before firing. The professor thought that such a habit related to outrunning the gamekeeper, and was intended to procure distance for the poacher rather than the bird.

Sachem stated that he had known a slow hunter once. He was a revolutionary veteran, used a revolutionary musket, and believed in revolutionary powder. He refused to do any thing different from what his fathers did, and abhorred double-barreled shotguns and percussion-caps as inventions of the devil. It was constantly, "General Washington did this," and "Our army did that," and his old head shook sadly at the innovations Young America was making. His ghost, with the revolutionary musket on its shoulder, had since been known to chase hunters, with breech-loaders, who were caught on his favorite ground after dark. "Old 1776" was great on wing-shooting, and could be seen at almost any time hobbling over the moor, firing away at snipe and waterfowl. He was one of those slow, deliberate cases, always taking snuff after the bird rose. There would be a glitter of fluttering wings as the game shot into air. Down would come the long musket, out would come the snuff-box, and the old soldier would go through the present, make ready, take snuff, take aim, and fire, all as coolly as if on parade. The old musket often hung fire from five to ten seconds, and the premonitory flash could be seen as the shaky flint clattered down on the pan. The veteran always patiently covered the bird until the charge got out. The recoil was tremendous, and the old man often went down before the bird; but such positions, he asserted, were taken voluntarily, as ones of rest. Some said that the gun had been known to kick him again after he was down."

Sachem's narration was here cut short by the dogs again pointing. This was followed by the usual bombardment, which over, the bag showed the magnificent aggregate of two chickens for the entire day's sport.

The prairie-chicken is now extinct in many of the Western States where it was once well known. Usually, during the first few years of settlement, it increases rapidly, and is often a nuisance to pioneer farmers. Perhaps, when the latter first settle in a country, a few covies may be seen; under the favorable influences of wheat and corn-fields, the dozens increase to thousands and cover the land. But with denser settlement come more guns, and, what is a far more destructive agent, trained dogs also. Under the first order of things, the farmer, with his musket, might kill enough for the home table. With double-barreled gun and keen-scented pointer, the sportsman and pot-hunter think nothing of fifty or sixty birds for a day's work. It seems almost impossible, under such a combination, for a covey to escape total annihilation.

We may suppose a couple of fair shots hunting over a dog in August, when the chickens lie close, and the year's broods are in their most delicate condition for the table. The pointer makes a stand before a fine covey hidden in the thick grass before him. The ready guns ask no delay, and, at the word, he flushes the chickens immediately under his nose. Each hunter takes those which rise before him, or on his side, and if four or less left cover at the first alarm, that number of gray-speckled forms the next moment are down in the grass, not to leave it again. If more rose, they are "marked," which means that their place of alighting is carefully noted, and, as the chicken has but a short flight, this task is easy. Meanwhile, the guns have been reloaded, the dog flushes others of the hiding birds, and so the sport goes on. The birds that get away are "marked down," and again found and flushed by the dog. Without this useful animal the chickens would multiply, despite any number of hunters. I have often seen covies go down in the grass but a few hundred yards away, yet have tramped through the spot dozens of times without raising a single bird. In twenty years this delicious game will probably be as much a thing of the past as is the Dodo of the Isle de France. At the period of our visit they were already gathering into their fall flocks, which sometimes number a hundred or more. In these they remain until St. Valentine recommends a separation. During the colder weather of winter they seek the protection of the timber, and may be seen of mornings on the trees and fences. They never roost there, however, but pass the night hidden in the adjacent grass.

The prairie-chicken's admirers are numerous, other animals beside man being willing to dine on its plump breast. We had an illustration of this in our first day's shooting. Sometimes when we fired, the report would attract to our vicinity wandering hawks, and we found that either instinct or previous experience teaches these fierce hunters of the air that in the vicinity of their fellow-hunter, man, wounded birds may be found. One wounded chicken, which fell near us, was seized by a hawk immediately.

As we passed one or two fields, indications of gophers appeared, their small mounds of earth covering the ground. In some counties these animals formerly destroyed crops to such an extent that the celebrated "Gopher Act" was passed. This gave a bounty of two dollars for each scalp, and under it many farms yielded more to the acre than ever before or since. One of these animals which we secured resembled in size and shape the Norway rat, and, in the softness and color of its coat, was not unlike a mole. The oddest thing was its earth-pouches—two open sacks, one on either side of its head, and capable of containing each a tablespoonful or more. These the gopher employs, in his subterranean researches, for the same purpose that his enemy, man, does a wheelbarrow. Packing them with dirt, the little fellow trudges gayly to the surface, and there cleverly dumps his load.

We reached town again, well pleased with our day's ride, and over our evening pipes discussed the results. Muggs thought our shot were too small. Sachem thought the birds were.

Colon was delighted with the new State, but believed that wing-shooting was not his *forte.* He would be more apt to hit a bird on the wing if he could only catch it roosting somewhere.

Gripe, at the other end of the room, was piling Republican doctrines upon a bearded Democratic heathen from the Western border.

CHAPTER V.

A TRIAL BY JUDGE LYNCH—HUNG FOR CONTEMPT OF COURT—QUAIL SHOOTING—HABITS OF THE BIRDS, AND MODE OF KILLING THEM—A RING OF QUAILS—THE EFFECTS OF A SEVERE WINTER—THE SNOW GOOSE.

A SHORT time after supper, Tenacious Gripe appeared with the mayor of the city, who wished to make the acquaintance of the Professor. The two august personages bowed to each other. It was the happiest moment in their respective lives, they declared. An invitation was extended us to delay our departure another day and try quail shooting. The citizens said the birds were unusually abundant, the previous winter having been mild and the summer long enough for two separate broods to be hatched, and the brush and river banks were swarming with them. As we were about to abandon the birds of the West and seek an acquaintance with its beasts, we decided, after a brief consultation, to accept the invitation and remain another day.

Among the persons present in the crowded office of the hotel, was a man from the southwestern part of the state who had lately been interested in a trial before the celebrated Judge Lynch. Sachem interviewed him, and reports his statement of the occurrence in the log book, as follows:

A stranger played me fur a fool,
An' threw the high, low, jack,
An' sold me the wuss piece of mule
That ever humped a back.

But that wer fair; I don't complain,
That I got beat in trade;
I don't sour on a fellow's gain,
When sich is honest made.

But wust wer this, he stole the mule,
An' I were bilked complete;
Such thieves, we hossmen makes a rule
To lift 'em from their feet.

We started arter that 'ere pup,
An' took the judge along,
For fear, with all our dander up,
We might do somethin' wrong.

We caught him under twenty miles,
An tried him under trees;
The judge he passed around the "smiles,"
As sort o' jury fees.

"Pris'ner," says judge, "now say your say,
An' make it short an' sweet,
An', while yer at it, kneel and pray,
For Death yer can not cheat.

No man shall hang, by this 'ere court,
Exceptin' on the square;
There's time fur speech, if so it's short,
But none to chew or swear."

JUDGE LYNCH—HIS COURT.

JUDGE AND JURY. SHERIFF. ATTORNEY. LOAFER. CLERK. DEPUTY SHERIFF.

Then judge, he frowned an awful frown,
An' snapped the sentence short:

" Jones, twitch the rope, an' write this down,
'Hung for contempt of court.' "

An' then the thievin' rascal cursed,
An' threw his life away,
He said, "Just pony out your worst,
Your best would be foul play."

Then judge he frowned an awful frown,
An' snapped this sentence short,
"Jones, twitch the rope, an' write this down,
Hung for contempt of court!"

Sharp 8 next morning saw us on the road leading east of town, the two dogs with us, and a young one additional, the property of a resident sportsman. Our last acquisition joined us on the run, and kept on it all day, going over the ground with the speed of a greyhound, his fine nose, however, giving him better success than his reckless pace would have indicated.

Three miles from town, or half way between it and Tecumseh, our party left the wagon, with direction for it to follow the road, while we scouted along on a parallel, following the river bank.

The Kaw stretched eastward, broad and shallow, with numerous sand bars, and along its edges grew the scarlet sumach and some stunted bushes, and between these and the corn a high, coarse bottom grass, with intervals at every hundred yards or so apart of a shorter variety, like that on a poor prairie. Among the bushes, there was no grass whatever, and yet the birds seemed indifferently to frequent one spot equally with another.

In less than ten minutes after leaving the wagon, all the dogs were pointing on a barren looking spot, thinly sprinkled with scrubby bushes not larger than jimson-weeds. They were several yards apart, so that each animal was clearly acting on his own responsibility.

If it puzzled us the day before to discover any signs of game under their noses, it certainly did so now. There was apparently no place of concealment for any object larger than a field-mouse. The bushes were wide apart, and the soil between was a loose sand. Around the roots of the scrubs, it is true, a few thin, wiry spears of grass struggled into existence, but these covered a space not larger than a man's hand, and it seemed preposterous to imagine that they could be capable of affording cover. That three dogs were pointing straight at three bushes was apparent, but we could see nothing in or about the latter calling for such attention.

Shamus, who had accompanied us, wished to know if the twigs were witch hazels, because, if so, three invisible old beldames might be taking a nap under them, after a midnight ride. "But, then," said Dobeen, "the dog's hairs don't stand on end as they always do in Ireland when they see ghosts and witches." We believe that our worthy cook was really disappointed in not discovering any stray broomsticks lying around. These, he afterward informed us, could not be made invisible, though their owners should take on airy shapes unrecognizable by mortal eyes.

Muggs had suggested urging the dogs in, but the party, wiser from yesterday's experience, desired a ground shot, if it could be secured. The Professor adjusted his lens, and decided to make a personal inspection around the roots of the bush immediately in front of him.

Carefully the sage bent over the suspicious spot, and almost fell backward as, with a whiz and a dart, half a dozen quails flew out, brushing his very nose. Instantly every bush sent forth its fugitives. A flash of feathered balls, and they were all gone. Such whizzing and whirring! it was as if those scraggy bushes were *mitrailleuses*, in quick succession discharging their loads.

Only one gun had gone off, but that so loudly that our ears rung for several seconds. Mr. Colon had accidentally rammed at least two, perhaps half a dozen, loads into one barrel, and the gun discharged with an aim of its own, the butt very low down. Two birds fell dead. But alas for our Nimrod! Colon stood with one hand on his stomach undecided whether that organ remained or not. On this point, however, he was fully re-assured at the supper-table that night, and in all our after experience, we never knew that gun to have the least opportunity for going off, except when at its owner's shoulder, and he perfectly ready for it.

The two birds were now submitted to the party for inspection. They were fine specimens of the American quail, more properly called by those versed in quailology, the Bob White. This bird is very plentiful throughout Kansas, and just before the shooting season commences, in September, will even frequent the gardens and alight on the houses of Topeka. They "lay close" before a dog, take flight into air with a quick, whirring dart, and their shooting deservedly ranks high. They are very rapid in their movements upon the ground, often running fifty or seventy-five yards before hiding. When this takes place, so closely do they huddle that it is seldom more than the upper bird that can be seen. "Green hunters" sometimes pause, trying to discover the rest of the covey before firing, and experience a great and sudden disgust when the single bird which they have disdained suddenly develops into a dozen flying ones.

We had an eventful days' sport, expending more ammunition than when among the chickens, and with more satisfactory results, as we brought in over two dozen birds. More than half of these were taken by Sachem at one lucky discharge. He saw a covey in the grass, huddled together as they generally are when not running. At these times they form a circle about as large in diameter as the hoop of a nail keg, with tails to the center and heads toward the outside. Fifteen quails would thus be a circle of fifteen heads, and a pail, could it be dropped over the covey, would cover them all. Not only is this an economy of warmth, there being no outsiders half of whose bodies must get chilled, but there is no blind side on which they can be approached, every portion of the circle having its full quota of eyes. Let skunk or fox, or other roamer through the grass, creep ever so stealthily, he will be seen and avoided by flight. Sachem aiming at the midst of such a ring, broke it up as effectually as Boutwell's discharge of bullion did that on Wall Street.

We have since found the frozen bodies of whole covies, which had gone to roost in a circle and been buried under such a heavy fall of snow that the birds could not force their way upward. Their habit is to remain in imprisonment, apparently waiting for the snow to melt before even making an effort for deliverance. Oftentimes it is then too late, a crust having formed above. A severe winter will sometimes completely exterminate the birds in certain localities.

During this first day of quail-shooting, we also saw for the first time flocks of the snow-goose. The Professor counted fifty birds on one sand bar. This variety, in its flight across the continent, apparently passes through but a narrow belt of country, being found, to the best of my knowledge, in but few of the states outside of Kansas.

Our return to the hotel was without accident, and our supper such as hungry hunters might well enjoy. After it was disposed of, we gathered around the ample stove in the hotel office, and lived over again the events of the day.

CHAPTER VI.

OFF FOR BUFFALO LAND—THE NAVIGATION OF THE KAW—FORT RILEY—THE CENTER-POST OF THE UNITED STATES—OUR PURCHASE OF HORSES—"LO" AS A SAVAGE AND AS A CITIZEN—GRIPE UNFOLDS THE INDIAN QUESTION—A BALLAD BY SACHEM, PRESENTING ANOTHER VIEW.

NEXT morning we said good-by to hospitable Topeka, and took up our westward way over the Pacific Railroad. An ever-repeated succession of valley and prairie stretched away on either hand. To the left the Kaw came down with far swifter current than it has in its course below, from its far-away source in Colorado. It might properly be called one of the eaves or water-spouts of the great Rocky Mountain water-shed. With a pitch of over five feet to the mile, its pace is here necessarily a rapid one, and when at freshet height the stream is like a mill-race for foam and fury.

At the junction of the Big Blue we found the old yet pretty town of Manhattan. To this point, in early times, water transit was once attempted. A boat of exceedingly light draught, one of these built to run on a heavy dew, being procured, freight was advertised for, and the navigation of the Kaw commenced. The one hundred miles or more to Manhattan was accomplished principally by means of the capstan, the boat being "warped" over the numberless shallows. This proved easier, of course—a trifle easier—than if she had made the trip on macadamized roads. If her stern had a comfortable depth of water it was seldom indeed, except when her bow was in the air in the process of pulling the boat over a sand bar. The scared catfish were obliged to retreat up stream, or hug close under the banks, to avoid obstructing navigation, and it is even hinted that more than one patriarch of the finny tribe had to be pried out of the way to make room for his new rival to pass.

Once at Manhattan, the steamboat line was suspended for the season, its captain and crew deciding they would rather walk back to the Missouri River than drag the vessel there. Soon afterward, the steamer was burned at her landing, and the Kaw has remained closed to commerce ever since.

About the same time, an enterprising Yankee advocated in the papers the straightening of the river, and providing it with a series of locks, making it a canal. As he had no money of his own with which to develop his ideas into results, and could command nobody's else for that purpose, the project failed in its very inception.

Fort Riley, four miles below Junction City, is claimed as the geographical center of the United States, the exact spot being marked by a post. What a rallying point that stick of wood will be for future generations! When the corner-stone of the National Capitol shall there be laid, the orator of the day can mount that post and exclaim, with eloquent significance, elsewhere impossible, "No north, no south, no east, no west!" and enthusiastic multitudes, there gathered from the four quarters of the continent, will hail the words as the key-note of the republic.

That spot of ground and that post are valuable. I hope a national subscription will be started to buy it. It is the only place on our continent which can ever be entirely free from local jealousies. There would be no possible argument for ever removing the capital. The Kaw could be converted into a magnificent canal, winding among picturesque hills past the base of the Capitol; and then, in case of war, should any hostile fleet ever ascend the rapid Missouri, it would be but necessary for our legislators to grasp the canal locks, and let the water out, to insure their perfect safety. Imagine the humiliation of a foreign naval hero arriving with his iron-clads opposite a muddy ditch, and finding it the only means of access to our capital!

A painful rumor has of late obtained circulation that a band of St. Louis ku-klux, yclept capital movers, intend stealing the pole and obliterating the hole. Let us hope, however, that it is without foundation.

Before leaving Topeka, the party had purchased horses for the trip, and consigned the precious load to a car, sending a note to General Anderson, superintendent, asking that they might be promptly and carefully forwarded to Hays City, our present objective point upon the plains.

The professor, bringing previous experience into requisition, selected a stout mustang—probably as tractable as those brutes ever become. He was warranted by the seller never to tire, and he never did, keeping the philosopher constantly on the alert to save neck and knees. It is the simple truth that, in all our acquaintance with him, that mustang never appeared in the least fatigued. After backing and shying all day, he would spend the night in kicking and stealing from the other horses.

Mr. Colon, by rare good fortune, obtained a beautiful animal, formerly known in Leavenworth as Iron Billy—a dark bay, with head and hair fine as a pointer's, limbs cut sharp, and joints of elastic. After carrying Mr. C. bravely for months, never tripping or failing, he was sold on our return to the then Secretary of State, who still owns him. More than once did Billy make his rider's arm ache from pulling at the curb, when the other horses were all knocked up by the rough day's riding. It was interesting to see him in pursuit of buffalo. He would often smell them when they were hidden in ravines, and we wholly unaware of their vicinity. Head and ears were erect in an instant, and, with nostrils expanded, forward he went, keeping eagerly in front at a peculiar prancing step which we called tiptoeing. Once in sight of the game, and the rider became a person of quite secondary importance. Billy said, as plainly as a horse could say any thing, "*I* am going to manage this thing; *you* stick on." And manage it he did. Not many moments, at the most, before he was at the quarters of the fleeing monsters, and nipping them mischievously with his teeth. I could always imagine him giving a downright horse-laugh, his big bright eyes sparkled so when the frightened bison, at the touch, gave a switch of his tail and a swerve of alarm, and plunged more wildly forward.

If the rider wished to shoot, he could do so; if not, content himself, as Mr. Colon usually did, with clinging to the saddle, and uttering numberless expostulatory but fruitless "whoa's."

Once on our trip Billy was loaned for the day to a gentleman who wished to examine a prospective coal mine. When barely out of sight of camp, Billy discovered a herd of buffalo, and, despite the vehement remonstrances of his rider, straightway charged it. The mine-seeker was no hunter, but a wise and thoroughly timid devotee of science in search of coal measures. A few moments, and the poor, frightened gentleman found himself in the midst of a surging mass of buffalo, his knees brushing their hairy sides, and their black horns glittering close around him, like an array of serried spears. He drew his knees into the saddle, and there, clinging like a monkey, lost his hat, his map of the mine, and his spectacles. He returned Billy as soon as he could get him back to camp, with expressions of gratitude that he had been allowed to escape with life, and never manifested the least desire to mount him again.

Sachem's purchase was a horse which had run unaccountably to legs. He was sixteen hands high, a trifle knock-kneed, and with a way of flinging the limbs out when put to his speed which, though it seemed awkward enough, yet got over the ground remarkably well. With the shambling gait of a camel, he had also the good qualities of one, and did his owner honest service.

Muggs bought a mule, partly because advised to do so by a plainsman, and partly because the rest of us took horses. With true British obstinacy he paid no attention to our expostulations, and the creature he obtained was as obstinate as himself. Poor Muggs! A mule may be good property in the hands of a plainsman, but was never intended to carry a Briton.

Semi-Colon had the auction purchase, and Dobeen selected a Mexican donkey, one of the toughest little animals that ever pulled a bit. He could excel a trained mule in the feat of dislodging his rider, and had a remarkable penchant for running over persons who by chance might be looking the other way. It seemed to be his constant study to take unexpected positions, or, as Sachem phrased it, to "strike an attitude."

My mount was a stout-built old mare, recommended to me as a solid beast, on the strength of which, and wishing to avoid experiments, I made purchase at once. I found her solid indeed. When on the gallop her feet came down with a shock which made my head vibrate, as if I had accidentally taken two steps instead of one, in descending a staircase.

Could the good people of Topeka have gotten us to ride out of their town upon our several animals, it would have given them a fair idea of a *mardi gras* cavalcade in New Orleans.

And so, our camp equipage and live stock following by freight, the express rolled us forward toward the great plains. So far along our route we had seen but few Indians, and those civilized specimens, such as straggle occasionally through the streets of Topeka. The Indian reservations in Kansas are at some distance apart, and their inhabitants frequently exchange visits. The few whom we had seen consisted of Osages, Kaws, Pottawatomies, and Sioux, all equally dirty, but the last affecting clothes more than the others, and eschewing paint. The members of this tribe, generally speaking, have good farms and are worth a handsome average per head. At the time of our visit they were expecting a half million dollars or so from Washington, and were soon to become American citizens. One privilege of this citizenship struck us as very peculiar. By the State law, as long as an Indian is simply an Indian, he can buy no whisky, and is thus cruelly debarred from the privilege of getting drunk, but once a voter, he can luxuriate in corn-juice and the calaboose, as well as his white brother. What a travesty upon American civilization and politics!

Muggs was prejudiced against the Osages, having been induced by one of them to invest in a bow and arrows, "for the Hinglish Museum, you know." On pulling for a trial shot, one end of the bow went further than the arrow, and the cord, warranted to be buffalo sinew, proved to be an oiled string.

Sachem declared that he had found Muggs returning the wreck to the Indian with the following speech: "O-sage, little was your wisdom to court thus the wrath of a Briton. Take with the two pieces this piece of my mind. That your noble form may be removed soon to the 'appy 'unting ground, where bow trades are not allowed, is the prayer of your patron, Muggs."

Mr. Colon asked Tenacious Gripe to explain the

UNNATURALIZED.

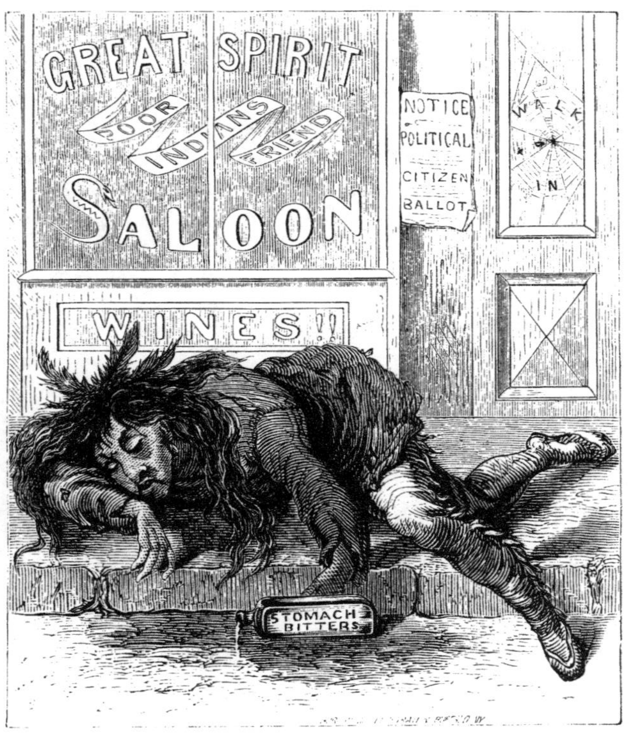

NATURALIZED.

condition of the Native Americans in Kansas. The orator kindly consented and thereupon discoursed as follows:

"The Indians of Kansas are divided into the wild and the tame. Both alike cover their nakedness with bright handkerchiefs, old shirts, military coats, and many-hued ribbons. The principal difference in point of dress is in the method of procuring it. Among those tribes which are at peace with the government, the white man robs the Indian; among the wild tribes the conditions are reversed—the Indian robs the white man. In the one case the contractors and agents carry off their half million dollars or thereabouts; in the other the savage bears away a quantity of old clothes and fresh scalps. As he finds it difficult to procure sufficient of the white man's justice to satisfy the cravings of his nature, he feeds it with what he can and whenever he can of revenge. Wise men tell us, gentlemen, that revenge is sweet and justice a dry morsel. All Indians beg when they get an opportunity. The tame ones, if they find it fruitless, divert themselves by selling worthless pieces of wood with strings attached, as bows. The wild ones, in a like predicament, relieve their tedium by whacking away at our ribs with bows that amount to something. The principles actuating both classes are alike. It is simply the application which causes difficulty—in the one case an appeal with bow and arrows to our pockets, in the other to our bodies.

"All our wars with these people, gentlemen, are a result of their political economy. They believe that the Great Spirit provided buffalo and other game for his red children. When the white man drives these away, they understand that he takes their place as a means of sustenance, and as they have lived upon the one, so they intend to do upon the other. If the buffalo attempts to evade his duty in the premises, they kill him and take his meat; if the white man, they kill him and take his hair."

Sachem produced a roll of dirty brown paper and said that he had studied the Indian question and found two sides to it. One he could give us in a nutshell, believing that the meat of the nut had often excited the spirit of war.

Where waters sung above the sand,
  And torrent forced its way,
Stretched out, disgusted with the land,
  A bearded miner lay,
Prepared to strike, with willing hand,
  Whatever lead would pay.

Echo of hoof on beaten ground
  Rung on the desert air,
Ringing a tune of gladsome sound
  To miner, watching there;
A paying lead, at last, he'd found—
  The vein a "man of hair."

An instant more, and at the ford
  A savage chief appeared;
The miner saw his goodly hoard,
  And tore his own good beard.
(You'll always find an ox is gored
  When sheep are to be sheared.)

Dog Town—The Happy Family at Home.

"You've riled that Brook"—An old Fable modernized.

And these the words the miner said:
"You've spoilt my drink, old fellow;
You've riled the brook, my brother red,
And, by your cheek so yellow,
To-night above your sandy bed
The prairie gale shall bellow.

"No relatives of mine are dead,
At least by Injun cunnin',
But many other hearts have bled,
And many eyes are runnin';
For blood and tears alike are shed,
When *you* go out a gunnin'.

"Some slumbrin' peaceful, first they knew,
They heard your horrid din—
Women as well as men you slew,
You bloody son of sin;
I mourn 'em all, revenge 'em too,
Through Adam they were kin."

This having said, the miner smart,
Drew bead upon the red man:
They're fond of beads—it touched his heart,
And Lo, behold, a dead man;
Upon Life's stage he'd played his part,
A gory sort of head man!

Two packs of goods lay on the ground;
Quoth miner, "Lawful spoil!
My lucky star at last has found
As good as gold and oil;
I kinder felt that fate was bound
To bless my honest toil.

"Such heathen have no lawful heirs—
I'll be the Probate Judge,
For though they kinder go in pairs,
Their love is all a fudge;
I'll 'ministrate on what he wears,
And leave his squaw my grudge."

CHAPTER VII.

ORIPE'S VIEWS OF INDIAN CHARACTER—THE DELAWARES' THE ISHMAELITES OF THE PLAINS—THE TERRITORY OF THE "LONG HORNS"—TEXANS AND THEIR CHARACTERISTICS—MUSHROOM ROCK—A VALUABLE DISCOVERY—FOOTPRINTS IN THE ROCK—THE PRIMEVAL PAUL AND VIRGINIA.

WE noticed many fine rivers rolling from the northward into the Kaw, which stream we found was known by that name only after receiving the Republican, at Junction City. Above that point, under the name of the Smoky Hill, it stretches far out across the plains, and into the eastern portion of Colorado. Along its desolate banks we afterward saw the sun rise and set upon many a weary and many a gorgeous day.

All the large tributaries of the Kansas river, consisting of the Big Blue, Republican, Solomon, and Saline, came in on our right. Upon our left, toward the South, only small creeks joined waters with the Kaw, the pitch of the great "divides" there being towards the Arkansas and its feeders, the Cottonwood and Neosho.

We had now fairly entered on the great Smcky Hill trail. Here Fremont marked out his path towards the Rocky Mountains and the Pacific, and on many of the high *buttes* we discovered the pillars of stone which he had set up as guides for emigrant trains, looking wonderfully like sentinels standing guard over the valleys beneath. Indeed we did at first take them for solitary herders, watching their cattle in some choice pasture out of sight.

Most of our party had expected to find Indians in promiscuous abundance over the entire State, and we were therefore surprised to see the country, after passing St. Mary's Mission, entirely free of them. Muggs asked Gripe if the American Indian was hostile to all nationalities alike, or simply to those who robbed him of his hunting-grounds. The orator replied as follows:

"Sir, the aborigine of the western plains cares not what color or flavor the fruit possesses which hangs from his roof tree. The cue of the Chinaman is equally as acceptable as hairs from the mane of the English lion. A red lock is as welcome as a black one, and disputes as to ownership usually result in a dead-lock. His abhorrence is a wig, which he considers a contrivance of the devil to cheat honest Indian industry. I would advise geologists on the plains to carry, along with their picks for breaking stones, a bottle of patent hair restorative. It is handy to have in one's pocket when his scalp is far on its way towards some Cheyenne war-pole. The scalping process, gentlemen, is the way in which savages levy and collect their poll-tax. Any person in search of romantic wigwams can have his wig warmed very thoroughly on the Arkansas or Texas borders. On the plains along the western border of Kansas, however, geologists can find a rich and comparatively safe field for exploration. It is doubtful if the savages ever wander there again.

"Of the Indian warrior on the plains we may well say, *requiescat in pace*, and may his pace be rapid towards either civilization or the happy hunting ground. History shows that his reaching the first has generally given him quick transit to the second. The white man's country has proved a spirit-land to Lo, whose noble soul seems to sink when the scalping-knife gathers any other rust than that of blood, and whose prophetic spirit takes flight at the prospect of exchanging boiled puppies and dirt for the white brother's pork and beans. Very often, however, it must be said, Lo's soul is gathered to his fathers by reason of its tabernacle being smitten too sorely by corn lightning."

As Gripe paused, the Professor took up the subject in a somewhat different strain:

"We have here in this State," remarked he, "a tribe which may well be called the Indian Ishmael. Its hand is and ever has been, since history took record of it, against its brethren, and its brethren's against it. I refer to the pitiful remnant of the once great Delawares. From the shores of the Atlantic they have steadily retreated before civilization, marking their path westward by constant conflicts with other races of red men. The nation in its eastern forests once numbered thousands of warriors. Now, three hundred miserable survivors are hastening to extinction by way of their Kansas reservation.

"A number of their best warriors have been employed as scouts by the government, when administering well merited chastisement to other murdering bands. The Delawares, I have often thought, are like blood-hounds on the track of the savages of the plains. They take fierce delight in scanning the ground for trails and the lines of the streams for camps. There is something strangely unnatural in the wild eyes of these Ishmaelites, as they lead the destroyers against their race, and assist in blotting it from the face of the continent. Themselves so nearly joined to the nations known only in history, it is like a plague-stricken man pressing eagerly forward to carry the curse, before he dies, to the remainder of his people."

The valleys of the Saline, Solomon, and Smoky Hill, as we passed them in rapid succession, seemed very rich and were already thickly dotted with houses. This is one of the best cattle regions of the state, and vast herds of the long-horned Texan breed covered the prairies. We were informed that they often graze throughout the entire winter. As early in the spring as the grass starts sufficiently along the trail from Texas to Kansas, the stock dealers of the former State commence moving their immense herds over it. The cattle are driven slowly forward, feeding as they come, and reach the vicinity of the Kansas railroads when the grass is in good condition for their summer fattening. As many as five hundred thousand head of these long horns have been brought into the State in a single season. Some are sold on arrival and others kept until fall, when the choicest beeves are shipped East for packing purposes, or into Illinois for corn feeding. The latter is the case when they are destined eventually for consumption in Eastern markets, grass-fed beef lacking the solid fatness of the corn-fed, and suffering more by long transportation.

This very important trade in cattle, when fully developed, will probably be about equally divided between southern and central Kansas, each of which possesses its peculiar advantages for the business. While the valley of the Arkansas has longer grass, and more of it, the dealers in the Kaw region claim that their "feed" is the most nutritious. My own opinion, carefully formed, is that both sections are about equally good, and that the whole of western Kansas, with Colorado, will yet become the greatest stock-raising region of the world. The climate is peculiarly favorable. Two seasons out of three, on an average, cattle and sheep can graze during the winter, without any other cover than that of the ravines and the timber along the creeks.

The herders who manage these large bodies of cattle are a distinctive and peculiar class. We saw numbers of them scurrying along over the country on their wild, lean mustangs, in appearance a species of centaur, half horse, half man, with immense rattling spurs, tanned skin, and dare-devil, almost ferocious faces. After an extensive acquaintance with the genus Texan, and with all due allowance for the better portion of it, I must say, as my deliberate judgment, that it embraces a larger number of murderers and desperadoes than can be found elsewhere in any civilized nation. A majority of these herders would think no more of snuffing out a life than of snuffing out a candle. Texas, in her rude solitude, formerly stretched protecting arms to the evil doers from other states, and to her these classes flocked. She offered them not a city but a whole empire of refuge.

Just beyond Brookville, two hundred miles from the eastern border of Kansas, our road commenced ascending the Harker Bluffs, a series of sandstone ridges bordering on the plains.

On our left, Mushroom Rock was pointed out to us, a huge table of stone poised on a solitary pillar, and strangely resembling the plant from which it is named. As the professor informed us, we were on the eastern shore of a once vast inland ocean, the bed of which now forms the plains. Sachem thought the rock might be a petrified toad-stool, on a scale with the gigantic toads which hopped around in the mud of that age of monsters. The professor thought it was fashioned by the waters, in their eddyings and washings.

Subsequent examinations showed this entire region to be one of remarkable interest to the geologist. A few miles east of Mushroom Rock, near Bavaria, as we learned from the conductor, human foot-prints had been discovered in the sandstone. The professor, who had long ascribed to man an earlier existence upon earth than that given him by geology, was greatly excited, and at his earnest request, when the down train was met, we returned upon it to Bavaria.

That place we found to consist of two buildings, each serving the double purpose of house and store,

MUSHROOM ROCK,

On Alum Creek, near Kansas Pacific R. R.—From a Photograph.

INDIAN ROCK, on Smoky Hill River, Kansas—From a Photograph.

the track running between them. Two sandstone blocks, each weighing several hundred pounds, lay in front of one of the stores, and there, sure enough, impressed clearly and deeply upon their surface were the tracks of human feet. They had been discovered by a Mr. J. B. Hamilton on the adjacent bluffs.

There was something weird and startling in this voice from those long-forgotten ages—ages no less remote than when the ridge we were standing upon was a portion of a lake shore. The man who trod those sands, the professor informed us, perished from the face of the earth countless ages before the oldest mummy was laid away in the caves of Egypt; and yet people looked at the shriveled Egyptian, and thought that they were holding converse with one who lived close upon the time of the oldest inhabitant. They wrested secrets from his tomb, and called them very ancient. And now this dweller beside the great lakes had lifted his feet out of the sand to kick the mummy from his pedestal of honor in the museum, as but a being of yesterday, in comparison with himself.

This discovery was soon afterward extensively noticed in the newspapers, and the specimens are now in the collection made by our party at Topeka. It is but fair to say that a difference of opinion exists in regard to these imprints. Many scientific men, among whom is Professor Cope, affirm that they must be the work of Indians long ago, as the age of the rock puts it beyond the era of man, while others attribute them to some lower order of animal, with a foot resembling the human one. For my own part, after careful examination, I accept our professor's theory, that the imprints are those of human feet. The surface of the stone has been decided by experts to be bent down, not chiseled out. Science not long ago ridiculed the primitive man, which it now accepts. It is not strange, therefore, that science should protest against its oldest inhabitant stepping out from ages in which it had hitherto for bidden him existence.

We also found on the rocks fine impressions of leaves, resembling those of the magnolia, and gathered a bushel of petrified walnuts and butternuts. There were no other indications whatever of trees, the whole country, as far as we could see, being a desolate prairie.

"Gentlemen," said the professor, "as surely as you stand on the shore of a great lake, which passed away in comparatively modern times, science stands on the brink of important revelations. We have here the evidence of the rocks that man existed on this earth when the vast level upon which you are about to enter was covered by its mass of water. The waves lapped against the Rocky Mountains on the west, and against the ridges on which you are standing, upon the east. From previous explorations, I can assure you that the buffalo now feed over a surface strewn with the remains of those monsters which inhabited the waters of the primitive world, and the grasses suck nutriment from the shells of centuries. Geology has held that man did not exist during the time of the great lakes. I assert that he did, gentlemen, and now an inhabitant of that period steps forward to confirm my position. This man walked barefooted, and yet the contour of one of the feet, so different in shape from that of any wild people's of the present day, shows that it had been confined by some stiff material, like our leather shoes. The appearance of the big toe is especially confirmatory of this. I would call your attention, gentlemen, to the block which contains companion impressions of the right and left foot. The latter is deep, and well defined, every toe being separate and perfect. The former is shallow, and spread out, with bulged-up ridges of stone between each toe. These are exactly the impressions your own feet would make, on such a shore to-day, were the sand under the right one to be of such a yielding nature that in moving you withdrew it quickly, and rested more heavily on the other, the material under which was firmer. Your right track would spread, the mud bulging up between the toes, and forcing them out of position, and the material nearly regaining its level, with a misshapen impression upon its surface.

"You will also perceive that the sand was already hardening into rock when our ancient friends walked over it. I use the plural because, if I may venture an opinion from this hasty examination, I should say the two tracks were those of a female, the single one that of a man. From the position of the blocks they were probably walking near each other at that precise time when the new rock was soft enough to receive an impression and hard enough to retain it. You will perceive that the surface of the stone is bent down into the cavities, as that of a loaf of half-raised bread would be should you press your hand into it."

Sachem thought that the couple might have been an ancient Paul and Virginia telling their love on the shores of the old-time lake.

The Professor continued: "You notice close beside the two imprints an oval, rather deep hole in the rock, precisely like that a boy often makes by whirling on one heel in the sand."

Sachem again interrupted: "Perhaps the maiden went through the fascinating evolution of revolving her body while her mind revolved the 'yes' or 'no' to her swain's question. It might be a refined way of telling her lover that she was well 'heeled,' and asking if he was."

The Professor very gravely replied: "In those days the world had not run to slang. If one of Noah's children had dared to address him with the modern salutation of 'governor,' the venerable patriarch would have flung his child overboard from the ark. Taking your view of the case, Mr. Sachem, the whirl in the sand, which gave the lover his answer, is telling us to-day that same old story. And the coquette of that remote period caused the tell-tale walk upon the sand, which has proved the greatest geological discovery of modern times. I believe that it will be followed up and sustained by others equally as important, all tending to date man's birth thousands of years anterior to the time geology has hitherto assigned him an existence upon earth."

We spent many hours of the night in getting the rocks to the depot for shipment to Topeka, the few inhabitants of Bavaria assisting us. Soon after a westward train came along, and we were again in motion toward the home of the buffalo.

Before we slept the Professor gave us the following information: The vast plateau lying east of the Rocky Mountains, and which we were now approaching, was once covered by a series of great fresh-water lakes. At an early period these must have been connected with the sea, their waters then being quite salty, as is abundantly demonstrated by the remains of marine shells. During the time of the continental elevation these lakes were raised above the sea level, and their size very much diminished. Over the new land thus created, and surrounding these beautiful sheets of water, spread a vegetation at once so beautiful and so rich in growth that earth has now absolutely nothing with which to compare it. Amid these lovely pastures roved large herds of elephants, with the mastodon, rhinoceros, horse, and elk, while the streams and lakes abounded with fish. But the drainage toward the distant ocean continued, the water area diminished, the hot winds of the dry land drank up what remained of the lakes, and, in process of time, lo! the great grass-covered plains that we wander over delightedly to-day. What folly to suppose that such a land, so peculiarly fitted for man's enjoyment, should remain, through a long period of time, tenanted simply by brutes, and be given up to the human race only after its delightful characteristics had been entirely removed.

CHAPTER VIII.

THE "GREAT AMERICAN DESERT"—ITS FOSSIL WEALTH—AN ILLUSION DISPELLED—FIRES ACCORDING TO NOVELS AND ACCORDING TO FACT—SENSATIONAL HEROES AND HEROINES—PRAIRIE DOGS AND THEIR HABITS—HAWK AND DOG AND HAWK AND CAT.

NEXT morning, as the first gray darts of dawn fell against our windows, Mr. Colon lifted up a sleepy head and gazed out. Then came that quick jerk into an upright position which one assumes when startled suddenly from a drowsy state to one of intense interest. The motion caused a similar one on the part of each of us, as if a sort of jumping-jack set of string nerves ran up our backs, and a man under the cars had pulled them all simultaneously.

We were on the great earth-ocean; upon either side, until striking against the shores of the horizon, the billows of buffalo-grass rolled away. It seemed as if the Mighty Ruler had looked upon these waters when the world was young, and said to them, "Ye waves, teeming with life, be ye earth, and remain in form as now, until the planet which bears you dissolves!" And so, frozen into stillness at the instant, what were then billows of water now stretch away billows of land into what seems to the traveler infinite distance, with the same long roll lapping against and upon distant *buttes* that the Atlantic has to-day in lashing its rock-ribbed coasts; and whenever man's busy industry cleaves asunder the surface, the depths, like those of ocean, give back their monsters and rare shells. Huge saurians, locked for a thousand centuries in their vice-like prison, rise up, not as of old to bask lazily in the sun, but to gape with huge jaws at the demons of lightning and steam rushing past, and to crack the stiff backs of savans with their forty feet of tail.

To the south of us, and distant several miles, was the line, scarcely visible, of the Smoky Hill, treeless and desolate; on the north, the upper Saline, equally barren. As difficult to distinguish as two brown threads dividing a brown carpet, they might have been easily overlooked, had we not known the streams were there, and, with the aid of our glasses, sought for their ill-defined banks.

A curve in the road brought us suddenly and sharply face to face with the sun, just rising in the far-away east, and flashing its ruddy light over the vast plain around us. Its bright red rim first appeared, followed almost immediately by its round face, for all the world like a jolly old jack tar, with his broad brim coming above deck. It reminded me on the instant of our brackish friend, Captain Walrus; and in imagination I dreamily pictured, as coming after him, with the broadening daylight, a troop of Alaskans, their sleds laden with blubber.

The air was singularly clear and bracing, producing an effect upon a pair of healthy lungs like that felt on first reaching the sea-beach from a residence inland. An illusion which had followed many of us from boyhood was utterly dissipated by the early dawn in this strange land. This was not the fact that the "great American desert" of our school-days is not a desert at all, for this we had known for years; it related to those floods of flame and stifling smoke with which sensational writers of western novels are wont to sweep, as with a besom of destruction, the whole of prairie-land once at least in every story. Young America, wasting uncounted gallons of midnight oil in the perusal of peppery tales of border life, little suspects how slight the foundation upon which his favorite author has reared the whole vast superstructure of thrilling adventure.

The scene of these heart-rending narratives is usually laid in a boundless plain covered with tall grass, and the *dramatis personæ* are an indefinite number of buffalo and Indians, a painfully definite one of emigrants, two persons unhappy enough to possess a beautiful daughter, and a lover still more unhappy in endeavoring to acquire title, a rascally half-breed burning to prevent the latter feat, and a rare old plainsman specially brought into existence to "sarcumvent" him.

At the most critical juncture the "waving sea of grass" usually takes fire, in an unaccountable manner—perhaps from the hot condition of the combatants, or the quantities of burning love and revenge which are recklessly scattered about. Multitudes of frightened buffalo and gay gazelles make the ground shake in getting out of the way, and the flames go to licking the clouds, while the emigrants

FIRE ON THE PLAINS, ACCORDING TO NOVELS.

FIRE ON THE PLAINS, AS IT IS.

go to licking the Indians. Although the fire can not be put out, one or the other, or possibly both, of the combatants are "put out" in short order.

Should the miserable parents succeed in getting their daughter safely through this peril, it is only because she is reserved for a further laceration of our feelings. The half-breed soon gets her, and the lover and rare old plainsman get on his track immediately afterward. And so on *ad libitum.*

We beg pardon for condensing into our sunrise reflections the material for a novel, such as has often run well through three hundred pages, and furnished with competencies half as many bill-posters. It is unpleasant to have one's traditionary heroes and heroines all knocked into pi before breakfast. It makes one crusty. Possibly, it may be their proper desert, but, if so, could be better digested after dinner.

The whole story would fail if the fire did, as novelists never like to have their heroines left out in the cold. But it is as impossible for flames as it is for human beings to exist on air alone. It is scarcely less so for them to feed, as they are supposed to do, on such scanty grass. The truth is, that what the bison, with his close-cropping teeth, is enabled to grow fat on, makes but poor material for a first-class conflagration.

The grass which covers the great plains of the Far West is more like brown moss than what its name implies. Perhaps as good an idea of it as is possible to any one who has never seen it, may be obtained by imagining a great buffalo robe covering the ground. The hair would be about the color and nearly the length of the grass, at the season in question. In the spring the plains are fresh and green, but the grass cures rapidly on the stalk, and before the end of July is brown and ripe. It will then burn readily, but the fire is like that eating along a carpet, and by no means terrifying to either man or brute. The only occasion when it could possibly prove dangerous is when it reaches, as it sometimes does, some of the narrow valleys where the tall grass of the bottom grows; but even then, a run of a hundred yards will take one to buffalo grass and safety. This latter fact we learned from actual experience, later on our trip.

What a wild land we were in! A few puffs of a locomotive had transferred us from civilization to solitude itself. This was the "great American desert" which so caught our boyish eyes, in the days of our school geography and the long ago. A mysterious land with its wonderful record of savages and scouts, battles and hunts. We had a vague idea then that a sphynx and half a score of pyramids were located somewhere upon it, the sand covering its whole surface, when not engaged in some sort of simoon performance above. No trains of camels, with wonderful patience and marvelous internal reservoirs of water, dragged their weary way along, it was true; yet that animal's first cousin, the American mule, was there in numbers, as hardy and as useful as the other. Many an eastern mother, in the days of the gold fever, took down her boys discarded atlas, and finding the space on the continent marked "Great

American Desert," followed with tearful eyes the course of the emigrant trains, and tried to fix the spot where the dear bones of her first-born lay bleaching.

As a people, we are better acquainted with the wastes of Egypt than with some parts of our own land. The plains have been considered the abode of hunger, thirst, and violence, and most of our party expected to meet these geniuses on the threshold of their domain, and, while Shamus should fight the first two with his skillet and camp-kettles to war against the third with rifle and hunting-knife.

But in the scene around us there was nothing terrifying in the least degree. The sun had risen with a clear highway before him, and no clouds to entangle his chariot wheels. He was mellow at this early hour, and scattered down his light and warmth liberally. Wherever the soil was turned up by the track, we discovered it to be strong and deep, and capable of producing abundant crops of resin weeds and sunflowers, which with farmers is a written certificate, in the "language of flowers," of good character.

We thundered through many thriving cities of prairie dogs, the inhabitants of which seemed all out of doors, and engaged in tail-bearing from house to house. The principal occupations of this animal appears to be two; first, barking like a squirrel, and second, jerking the caudal appendage, which operations synchronize with remarkable exactitude. One single cord seems to operate both extremities of the little body at once. It could no more open its mouth without twitching its tail, than a single-thread Jack could bow its head without lifting its legs. Those nearest would look pertly at us for a moment, and then dive head foremost into their holes. The tail would hardly disappear before the head would take its place and, peering out, scrutinize us with twinkling eyes, and chatter away in concert with its neighbors, with an effect which reminded me of a forest of monkeys suddenly disturbed.

Sachem declared that they must all be females, for no sooner had one been frightened into the house than it poked its head out again to see what was the matter. "That sex would risk life at any time to know what was up."

The professor, with a more practical turn, told us some of the quaint little animal's habits. "Why it is called a dog," said he, "I do not know. Neither in bark, form, or life, is there any resemblance. It is carnivorous, herbivorous, and abstemious from water, requiring no other fluids than those obtained by eating roots. Its villages are often far removed from water, and when tamed it never seems to desire the latter, though it may acquire a taste for milk. It partakes of meats and vegetables with apparently equal relish. It is easily captured by pouring two or three buckets of water down the hole, when it emerges looking somewhat like a half-drowned rat. The prairie dog is the head of the original 'happy family.' It was formerly affirmed, even in works of natural history, that a miniature evidence of the millennium existed in the home of this little animal. There the rattlesnake, the owl, and the dog were supposed to lie down together, and such is still the general belief. It was known that the bird and the reptile lived in these villages with the dog, and science set them down as honored guests, instead of robbers and murderers, as they really are."

On our trip we frequently killed snakes in these villages which were distended with dogs recently swallowed. The owls feed on the younger members of the household, and the old dogs, except when lingering for love of their young, are not long in abandoning a habitation when snakes and owls take possession of it. The latter having two votes, and the owner but one (female suffrage not being acknowledged among the brutes), it is a "happy family," on democratic principles of the strictest sort.

We have also repeatedly noticed the dogs busily engaged in filling up a hole quite to the mouth with dirt, and have been led to believe that in this manner they occasionally revenge themselves upon their enemies, perhaps when the latter are gorged with tender puppies, by burying them alive. An old scout once told us that this filling up process occurred whenever one of their community was dead in his house, but as the statement was only conjectural, we prefer the other theory.

While we were this day steaming through one village an incident occurred showing that these animals have yet another active enemy. Startled by the cars, the dogs were scampering in all directions, when a powerful chicken-hawk shot down among them with such wonderful rapidity of flight that his shadow, which fell like that from a flying fragment of cloud, scarcely seemed to reach the earth before him. Some hundreds of the little brown fellows were running for dear life, and plunging wildly into their holes without any manifestations of their usual curiosity. The hawk's shadow fell on one fat, burgher-like dog, perhaps the mayor of the town, and in an instant the robber of the air was over him and the talons fastened in his back. Then the bird of prey beat heavily with its pinions, rising a few feet, but, finding the prize too heavy, came down. He was evidently frightened at the noise of the cars and we hoped the prisoner would escape. But the bird, clutching firmly for an instant the animal in its talons, drew back his head to give force to the blow, and down clashed the hooked beak into one of the victim's eyes. A sharp pull, and the eyeball was plucked out. Back went the beak a second time, and the remaining eye was torn from its socket, and the sightless body was then left squirming on the ground, while the hawk flew hastily away a short distance, evidently to return when we had passed on. It was pitiful to see the dog raise up on its haunches and for an instant sit facing us with its empty sockets, then make two or three short runs to find a path, in its sudden darkness, to some hole of refuge, but fruitlessly, of course.

A few days afterward, at Hays City, we witnessed an affair in which the air-pirate got worsted. While sitting before the office of the village doctor, a powerful hawk pounced upon his favorite kitten, which lay asleep on the grass, and started off with it. The two had reached an elevation of fifty feet, when puss recovered from her surprise and went to work for liberty. She had always been especially addicted to dining on birds, and the sensation of being carried off by one excited the feline mind to astonishment and wrath. Twisting herself like a weasel her claws came uppermost, and to our straining gaze there was a sight presented very much as if a feather-bed had been ripped open. The surprised hawk had evidently received new light on the subject; it let go on the instant, and went off with the appearance of a badly plucked goose, while the cat came safely to earth and sought the nearest way home.

CHAPTER IX.

WE SEE BUFFALO—ARRIVAL AT HAYS—GENERAL SHERIDAN AT THE FORT—INDIAN MURDERS—BLOOD-CHRISTENING OF THE PACIFIC RAILROAD—SURPRISED BY A BUFFALO HERD—A BUFFALO BULL IN A QUANDARY—GENTLE ZEPHYRS—HOW A CIRCUS WENT OFF—BOLOGNA TO LEAN ON—A CALL UPON SHERIDAN.

AS we passed out of the dog village, the engine gave several short, sharp whistles, and numberless heads were at once thrust out to ascertain the cause. "Buffalo!" was the cry, and with this there was a rush to the windows for a view of the noblest of American game. Even sleepy elderly gentlemen jostled rudely, and Sachem forgot his liver so far as to crowd into a favorable position beside a young woman.

"There they go!" "Oh, my, what monsters!" "What beards!" "What horns!" "Beats a steeple-chase!" "Uncanny beasts, lookin' and gangin' like Nick!" "Sure, they're going home from a divil's wake!" and similar ejaculations filled the car, as they do a race-stand when the horses are off. Two huge bulls had crossed just ahead of the engine, and one of them, apparently deeming escape impossible, was standing at bay close to the track, head down for a charge. He was furious with terror, the hissing steam and cow-catcher having been close at his heels for a hundred yards. As we flew past he was immediately under our windows, and we were obliged to look down to get a view of his immense body, with the back curving up gradually from the tail into an uncouth hump over the fore shoulders.

These two solitary old fellows were the only buffalo we saw from the train, the herds at large having not yet commenced their southern journey. At certain seasons, however, they cover the plains on each side of the road for fifty or sixty miles in countless multitudes. These wild cattle of Uncle Samuel's, if called upon, could supply the whole Yankee nation with meat for an indefinite period.

About noon we arrived at Hays City, two hundred and eighty miles from the eastern border of the State, and eighty miles out upon the plains. A stream tolerably well timbered, known as Big Creek, runs along the southern edge of the town, and just across it lies Fort Hays, town and fort being less than a mile apart.

The post possessed considerable military importance, being the base of operations for the Indian country. We found Sheridan there, an officer who won his fame gallantly and on the gallop. During the summer our red brethren had been gathering a harvest of scalps, and, in return, our army was now preparing to gather in the gentle savage.

We had read accounts in the newspapers, some time before, of the capture of Fort Wallace and of attacks on military posts. Such stories were not only untrue, but exceedingly ridiculous as well. Lo is not sound on the assault question. His chivalrous soul warms, however, when some forlorn Fenian, with spade on shoulder and thoughts far off with Biddy in Erin's Isle, crosses his vision. Being satisfied that Patrick has no arms, his only defense being utter harmlessness, and well knowing that the sight of a painted skin, rendered sleek by boiled dog's meat, will make him frantic with terror, the soul of the noble savage expands. No more shall the spade, held so jauntily, throw Kansas soil on the bed of the Pacific Railroad; and the scalp, yet tingling with the boiling of incipient Fenian revolutions underneath, on the pole of a distant wigwam will soon gladden the eyes of the traditionally beautiful Indian bride, as with dirty hands she throws tender puppies into the pot for her warrior's feast. The savage hand, crimson since childhood, descends with defiant ring upon the tawny breast, and, with a cry of, "Me big Indian, ha, whoop!" down sweeps Lo upon the defenseless Hibernian. A startled stare, a shriek of wild agony, a hurried prayer to "our Mary mother," and Erin's son christens those far-off points of the Pacific Railroad with his blood. A rapid circle of hunting-knife and the scalp is lifted, a few twangs of the bow fills the body with arrows, there is a rapid vault into the saddle, and a mutilated corpse, with feathered tips, like pins in a cushion, dotting its surface, alone remains to tell the tale of horror.

Blood had been every-where on the railroad, which reached across the plains like a steel serpent spotted with red. There was now a cessation of hostilities, and Indian agents were reported to be on the way from Washington to pacify the tribes. As they had been a long time in coming, the inference was irresistible that the popping of champagne corks was a

BUREAU OF ILLUSTRATION BUFFALO, N.Y.

" And Erin's son christens those far-off points of the Pacific Railroad with his blood."

much more pleasant experience than that of Indian guns would have been. The harvest of scalps had reached high noon some time before. Far off, south of the Arkansas, the savages had their home, and from thence, like baleful will-o'-the-wisps, they would suddenly flash out, and then flash back when pursued, and be lost in those remote regions. Lately, United States troops have been so placed that the Indian villages may be struck, if necessary, and retaliation had; and this, together with the pacificatory efforts of the Quaker agents, is doing much to bring about a condition of things which promises permanent peace.

Here our party was at Hays, the objective point of our journey, and our base of operations against the treasures of the past and present, which alike covered the country around. This little town is in the midst of the great buffalo range. Away upon every side of it stretch those vast plains where the short, crisp grass curls to the ridges, like an African's kinky hair to his skull. Bison and wild horse, antelope and wolf, for weeks were now to be our neighbors, appearing and vanishing over the great expanse like large and small piratical crafts on an ocean. We were kindly received at the Big Creek Land Company's office, on the outskirts of the town, and there deposited our guns and baggage. Our horses were expected on the morrow.

Twilight found us, after a busy afternoon, sitting around the office door, with that tired feeling which a traveler has when mind and body are equally exhausted. Our very tongues were silent, those useful members having wagged until even they were grateful for the rest. The hour of dusk, of all others, is the time for musing, and almost involuntarily our minds wandered back a twelve-month, when the plains were a solitude. No railroad, no houses, no tokens of civilization save only a few solitary posts, garrisoned with corporal's guards, and surrounded by red fiends thirsty for blood. Such was the picture then; now, the clangor of a city echoed through Big Creek Valley.

While wondering at the change, away on the hills to our right there rose a thundering tread, like the marching of a mighty multitude. Shamus, who sat directly facing the hill, saw something which chilled the Dobeen blood, and caused that noble Irishman to plunge behind us. Mr. Colon, who had given a startled turn of the head over his right shoulder, exclaimed, "Bless me, what's that?" The glance of Muggs froze that Briton so completely that he failed to tell us of ever having seen a more "hextraordinary thing in Hingland." I am in doubt whether even our grave professor did not imagine for the moment that the mammalian age was taking a tilt at us.

Gathering twilight had magnified what in broad day would have been an apparition sufficiently startling to any new arrival in Buffalo Land. A long line of black, shaggy forms was standing on the crest and looking down upon us. It had come forward like the rush of a hungry wave, and now remained as one uplifted, dark and motionless. In bold relief against the horizon stood an array of colossal figures, all bristling with sharp points, which at first sight seemed lances, but at the second resolved into horns. Then it dawned upon our minds that a herd of the great American bison stood before us. What a grateful reduction of lumps in more than one throat, and how the air ran riot in lately congealed lungs!

Dobeen declared he thought the professor's "ghosts of the centuries" had been looking down upon us.

One old fellow, evidently a leader in Buffalo Land, with long patriarchial beard and shaggy forehead, remained in front, his head upraised. His whole attitude bespoke intense astonishment. For years this had been their favorite path between Arkansas and the Platte. Big Creek's green valley had given succulent grasses to old and young of the bison tribe from time immemorial. Every hollow had its traditions of fierce wolf fights and Indian ambuscades, and many a stout bull could remember the exact spot where his charge had rescued a mother and her young from the hungry teeth of starving timber wolves. Every wallow, tree, and sheltering ravine were sacred in the traditions of Buffalo Land. The petrified bones of ancestors who fell to sleep there a thousand years before testified to purity of bison blood and pedigree.

Now all this was changed. Rushing toward their loved valley, they found themselves in the suburbs of a town. Yells of red man and wolf were never so horrible as that of the demon flashing along the valley's bed. A great iron path lay at their feet, barring them back into the wilderness. Slowly the shaggy monarch shook his head, as if in doubt whether this were a vision or not; then whirling suddenly, perhaps indignantly, he turned away and disappeared behind the ridge, and the bison multitude followed.

Our horses arrived the next morning all safe, excepting a few skin bruises, the steed Cynocephalus, however, being a trifle stiffer than usual, from the motion of the cars. When they were trotted out for inspection, by some hostlers whom we had hired that morning for our trip, the inhabitants must have considered the sight the next best thing to a circus.

Apropos of circuses, we learned that one had exhibited for the first and only time on the plains a few months before. In that country, dear reader, Æolus has a habit of loafing around with some of his sacks in which young whirlwinds are put up ready for use. One of these is liable to be shaken out at any moment, and the first intimation afforded you that the spirit which feeds on trees and fences is loose, is when it snatches your hat, and begins flinging dust and pebbles in your eyes. But to return to our circus performance. For awhile all passed off admirably. The big tent swallowed the multitude, and it in turn swallowed the jokes of the clown, older, of course, than himself. In the customary little tent the living skeleton embodied Sidney Smith's wish and sat cooling in his bones, while the learned pig and monkey danced to the melodious accompaniment of the hand-organ.

Suddenly there was a clatter of poles, and two canvass clouds flew out of sight like balloons. The living skeleton found himself on a distant ridge, with the wind whistling among his ribs, while the monkey

GENTLE ZEPHYRS—"GOING OFF WITHOUT A DRAWBACK."

performed somersaults which would have astonished the original Cynocephalus. The pig meanwhile found refuge behind the organ, which the hurricane, with a better ear for music than man, refused to turn.

"Mademoiselle Zavenowski, the beautiful leading equestrienne of the world," just preparing to jump through a hoop, went through her own with a whirl, and stood upon the plains feeding the hungry storm with her charms. The graceful young rider, lately perforating hearts with the kisses she flung at them, in a trice had become a maiden of fifty, noticeably the worse for wear.

An eye-witness, in describing the scene to us, said the circus went off without a single drawback. It was as if a ton of gunpowder had been fired under the ring. Just as the clown was rubbing his leg, as the result of calling the sensitive ring-master a fool (a sham suffering, though for truth's sake), there was a sharp crack, and the establishment dissolved. High in air went hats and bonnets, like fragments shot out of a volcano. The spirits of zephyr-land carried off uncounted hundreds of tiles, both military and civil, and we desire to place it upon record that should a future missionary, in some remote northern tribe, find traditions of a time when the sky rained hats, they may all be accounted for on purely scientific grounds.

Much property was lost, but no lives. The immediate results were a bankrupt showman and a run on liniments and sticking-plaster.

Our first hunt was to be on the Saline, which comes down from the west about fifteen miles north of Hays City.

Before starting, we carefully overhauled our entire outfit. For a long, busy day nothing was thought of save the cleaning of guns, the oiling of straps, and the examination of saddles, with sundry additions to wardrobe and larder. Shamus became a mighty man among grocery-keepers, and could scarcely have been more popular had he been an Indian supply agent. The inventory which he gave us of his purchases comprised twelve cans of condensed milk, with coffee, tea, and sugar, in proportion; several pounds each of butter, bacon, and crackers; a few loaves of bread, two sacks of flour, some pickles, and a sufficient number of tin-plates, cups, and spoons. To these he subsequently added a half-dozen hams and something like fifty yards of Bologna sausage, which he told us were for use when we should tire of fresh meat. Sachem entered protest, declaring that sausage and ham, in a country full of game, reflected upon us.

Of course, we found use for every item of the above, and especially for the Bologna. If one can feel satisfied in his own mind as to what portion of the brute creation is entering into him, a half-yard of Bologna, tied to the saddle, stays the stomach wonderfully on an all day's ride. It is so handy to reach it, while trotting along, and with one's hunting-knife cut off a few inches for immediate consumption. Semi-Colon, however, who was a youth of delicate stomach, sickened on his ration one day, because he found some-

"LOOKED LIKE THE END OF A TAIL."

THE RARE OLD PLAINSMAN OF THE NOVELS.

thing in it which, he said, looked like the end of a tail. It is a debatable question, to my mind, whether Satan, among his many ways of entering into man, does not occasionally do so in the folds of Bologna sausage. Certain it is that, after such repast, one often feels like Old Nick, and woe be to the man at any time who is at all dyspeptic. All the forces of one's gastric juices may then prove insufficient to wage successful battle with the evil genius which rends him.

Our outfit, as regards transportation, consisted of the animals heretofore mentioned, and two teams which we hired at Hays, for the baggage and commissary supplies.

The evening before our departure we rode over to the fort and called upon General Sheridan. "Little Phil" had pitched his camp on the bank of Big Creek, a short distance below the fort, preferring a soldier's life in the tent to the more comfortable officer's quarters. This we thought eminently characteristic of the man. He is an accumulation of tremendous energy in small compass, a sort of embodied nitro-glycerine, but dangerous only to his enemies. Famous principally as a cavalry leader, because Providence flung him into the saddle and started him off at a gallop, had his destiny been infantry, he would have led it to victory on the run. And now, officer after officer having got sadly tangled in the Indian web, which was weaving its strong threads over so fair a portion of our land, Sheridan was sent forward to cut his way through it.

The camp was a pretty picture with its line of

white tents, the timber along the creek for a background, and the solemn, apparently illimitable plains stretching away to the horizon in front. Taken altogether, it looked more like the comfortable nooning spot of a cavalry scout than the quarters of a famous General. Our chieftain stood in front of the center tent, with a few staff-officers lounging near by, his short, thick-set figure and firm head giving us somehow the idea of a small, sinewy lion.

We found the General thoroughly conversant with the difficult task to which he had been called. "Place the Indians on reservations," he said, "under their own chiefs, with an honest white superintendency. Let the civil law reign on the reservation, military law away from it, every Indian found by the troops off from his proper limits to be treated as an outlaw." It seemed to me that in a few brief sentences this mapped out a successful Indian policy, part of which indeed has since been adopted, and the remainder may yet be.

When speaking of late savageries on the plains the eyes of "Little Phil" glittered wickedly. In one case, on Spillman's Creek, a band of Cheyennes had thrust a rusty sword into the body of a woman with child, piercing alike mother and offspring, and, giving it a fiendish twist, left the weapon in her body, the poor woman being found by our soldiers yet living.

"I believe it possible," said Sheridan, "at once and forever to stop these terrible crimes." As he spoke, however, we saw what he apparently did not, a long string of red tape, of which one end was pinned to his official coat-tail, while the other remained in the hands of the Department at Washington. Soon after, as Sheridan pushed forward, the Washington end twitched vigorously. He managed, however, with his right arm, Custer, to deal a sledge-hammer blow, which broke to fragments the Cheyenne Black-kettle and his band. Whether or not that band had been guilty of the recent murders, the property of the slain was found in their possession, and the terrible punishment caused the residue of the tribe to sue for peace. It was the first time for years that the war spirit had placed any horrors at their doors, and that one terrible lesson prepared the savage mind for the advent of peace commissioners.

Our brief conference ended, the General bade us good day, and wished us a pleasant experience. Scarcely had we got beyond his tents, however, when we were overtaken by a decidedly unpleasant one. On their way to water, a troop of mules stampeded, and passing us in a cloud of dust, our brutes took bits in their teeth, and joined company. Happily, the run was a short one to the creek, where those of us who had not fallen off before managed to do so then. Poor Gripe was the only person injured, suffering the fracture of a rib, which necessitated his return to Topeka, so that we did not see him again until some months afterward, when we met him on the Solomon.

CHAPTER X.

HAYS CITY BY LAMP-LIGHT—THE SANTA FE TRADE—BULL-WHACKERS—MEXICANS—SABBATH ON THE PLAINS—THE DARK AGES—WILD BILL AND BUFFALO BILL—OFF FOR THE SALINE—DOBEEN'S GHOST-STORY—AN ADVENTURE WITH INDIANS—MEXICAN CANNONADE—A RUNAWAY.

HAYS CITY by lamp-light was remarkably lively and not very moral. The streets blazed with the reflection from saloons, and a glance within showed floors crowded with dancers, the gaily dressed women striving to hide with ribbons and paint the terrible lines which that grim artist, Dissipation, loves to draw upon such faces. With a heartless humor he daubs the noses of the sterner sex a cherry red, but paints under the once bright eyes of woman a shade dark as the night in the cave of despair. To the music of violin and stamping of feet, the dance went on, and we saw in the giddy maze old men who must have been pirouetting on the very edge of their graves.

Being then the depot for the great Santa Fe trade, the town was crowded with Mexicans and speculators. Large warehouses along the track were stored with wool awaiting shipment east, and with merchandise to be taken back with the returning wagons. These latter are of immense size, and, from this circumstance, are sometimes called "prairie schooners;" and, in truth, when a train of them is winding its way over the plains, the white covers flecking its surface like sails, the sight is not unlike a fleet coming into port. Oxen and mules are both used. When the former, the drivers rejoice in the title of "bull-whackers," and the crack of their whips, as loud as the report of a rifle, is something tremendous.

On the day of our arrival at Hays City, one of these festive individuals noticed Dobeen gazing, with open mouth, and back towards him, at some object across the street, and took the opportunity to crack his lash within an inch of the Irishman's spine. The effect was ludicrous; Shamus came in on the run to have a ball extracted from his back!

These Mexicans who come through with the ox-trains are a very degraded race, dark, dirty, and dismal. In appearance, they much resemble animated bundles of rags, walking off with heads of charcoal. Personal bravery is not one of their striking characteristics; indeed, they often run away when to stand still would seem to an American the only safe course possible. We were desirous of sending back to Hays City some of the proceeds of our excursion for shipment to friends at St. Louis and Chicago, and therefore hired two of the Mexican teamsters to go as far as the Saline, and return with the fruits of our prowess. For this service, which would occupy about four days, they were to receive twenty-five dollars each.

The morrow was Sunday, and came to us, as nine-tenths of the mornings on the plains did afterward, clear and bracing. Compared with the previous evening, the little town was very quiet. There was no stir in the streets, although later in the morning a few of the last night's carousers came out of doors, rubbing their sleepy eyes, and slunk around town for the remainder of the day. All nature was calm and beautiful; it almost seemed as if we might hear the chime of Sabbath bells float to us from somewhere in the depths around.

One of our sea legends recites that ship wrecked bells, fallen from the society of men to that of mermaids, are straightway hung on coral steeples, where, when storms roar around the rocks above, they toll for the deaths of the mariners. Was it impossible, we mused, that ancient mariners, with whole cargoes of bells, went down on this inland sea centuries before Rome howled? The earth around us might be as full of musical tongues as of saurians, and only awaiting the savan's spade and sympathetic touch to give their dumb eloquence voice. If the people of those days were navigators, surely they might also have been men of metal. In the far-away past existed numerous arts which baffle modern ingenuity. Stones were lifted at sight of which our engineers stand dismayed. Bodies were embalmed with a skill and perfection which our medical faculty admire, but have scarcely even essayed to imitate. Is it impossible that vessels plowed this ancient ocean with a speed which would have left our Cunarders out of sight? If human spirits freed from earth take cognizance of following generations, how those old captains must have laughed when Fulton boarded his wheezing experiment to paddle up the Hudson! And if our doctor's Darwinian-Pythagorean theory were correct, Fulton's spirit might have brought the crude idea from some ancient stoker.

But while we were thus speculating and giving free reins to Fancy's most erratic moods, the chaplain arrived from the fort, and mounting the freight platform, read the Episcopal morning service. A crowd gathered around, and a voice from the past whispering in their ears, a few bowed their heads during prayer. A drunkard went brawling by, with a sidelong glance and the leering look of eyes whose watery lids seemed making vain efforts to quench the fiery balls. How it grated on one's feelings! In a land so eloquent with voices of the mighty past, it seemed as if even instinct would cause the knee to bow in homage before its Maker.

Monday was our day of final preparation, and we commenced it by making the acquaintance of those two celebrated characters, Wild Bill and Buffalo Bill, or, more correctly, William Hickock and William Cody. The former was acting as sheriff of the town, and the latter we engaged as our guide to the Saline.

Wild Bill made his *entree* into one court of the temple of fame some years since through Harper's Magazine. Since then his name has become a household word to residents along the Kansas frontier. We found him very quiet and gentlemanly, and not at all the reckless fellow we had supposed. His form won our admiration—the shoulders of a Hercules with the waist of a girl. Much has been written about Wild Bill that is pure fiction. I do not believe, for example, that he could hit a nickel across the street with a pistol-ball, any more than an Indian could do so with an arrow. These feats belong to romance. Bill is wonderfully handy with his pistols, however. He then carried two of them, and while we were at Hays snuffed a man's life out with one; but this was done in his capacity of officer. Two rowdies devoted their energies to brewing a riot, and defied arrest until, at Bill's first shot, one fell dead, and the other threw up his arms in token of submission. During his life time Bill has probably killed his baker's dozen of men, but he has never, I believe, been known as the aggressor. To the people of Hays he was a valuable officer, making arrests when and where none other dare attempt it. His power lies in the wonderful quickness with which he draws a pistol and takes his aim. These first shots, however, can not always last. "They that take the sword shall perish with the sword;" and living as he does by the pistol, Bill will certainly die by it, unless he abandons the frontier.

Only a short time after we left Hays two soldiers attempted his life. Attacked unexpectedly, Bill was knocked down and the muzzle of a musket placed against his forehead, but before it could be discharged the ready pistol was drawn and the two soldiers fell down, one dead, the other badly wounded. Their companions clamored for revenge, and Bill changed his base. He afterward became marshal of the town of Abilene, where he signalized himself by carrying a refractory councilman on his shoulders to

BUFFALO BILL—FROM A PHOTOGRAPH.

WILD BILL—FROM A PHOTOGRAPH.

the council-chamber. A few months later some drunken Texans attempted a riot, and one of them, a noted gambler, commenced firing on the marshal. The latter returned the fire, shooting not only the gambler, but one of his own friends, who, in the gloom of the evening, was hurrying to his aid. Bill paid the expenses of the latter's funeral, which on the frontier is considered the proper and delicate way of consoling the widow whenever such little accidents occur.

The Professor took occasion, before parting with Wild William, to administer some excellent advice, urging him especially, if he wished to die in his bed, to abandon the pistol and seize upon the plow-share. His reputation as Union scout, guide for the Indian country, and sheriff of frontier towns, our leader said, was a sufficient competency of fame to justify his retirement upon it. In this opinon the public will certainly coincide.

Buffalo Bill was to be our guide. He informed us that Wild Bill was his cousin. Cody is spare and wiry in figure, admirably versed in plain lore, and altogether the best guide I ever saw. The mysterious plain is a book that he knows by heart. He crossed it twice as teamster, while a mere boy, and has spent the greater part of his life on it since. He led us over its surface on starless nights, when the shadow of the blackness above hid our horses and the earth, and though many a time with no trail to follow and on the very mid-ocean of the expanse, he never made a failure. Buffalo Bill has since figured in one of Buntline's Indian romances. We award

8

him the credit of being a good scout and most excellent guide; but the fact that he can slaughter buffalo is by no means remarkable, since the American bison is dangerous game only to amateurs.

We were off early on Tuesday morning for the Saline, our course toward which lay before us a little west of north, the citizens turning out to see us start. We had just parted from Gripe, who went East on the first train to get his ribs healed. "To think, gentlemen," said he, "that I should have escaped rebel bullets and Indian atrocities, only to have my ribs cracked at last by a stampede of mules!" Poor Gripe's farewell reminded me strongly of the old saying about the ruling passion strong in death. As he stood on the platform, with one hand against his aching side, he could not refrain from waving a courtly adieu with the other, and bowing himself from our presence, into the car, as if leaving the stage after a political speech.

We were sorry to lose our friend, and this, together with the thought of the weeks of uncertainties and anxieties which lay before us, made our exit from Hays rather a solemn affair. Even Tammany Sachem's face was ironed out so completely that not a smile wrinkled it. Dobeen had loaded one wagon with culinary weapons, and now sat among his pots and pans, evidently ill at ease and wishing himself doing any thing else rather than about to plunge further into the wilderness.

When about to mount Cynocephalus, Semi's feelings were wounded by a depraved urchin who suggested, "You'd better fust knock that fly off, Boss.

Both on ye 'll be too much for the hoss!" Fortunately, perhaps, for our feelings, the remainder of the inhabitants were so civil that further criticisms on our outfit, though they may have been ripe at their tongues' end, were carefully repressed.

Moving out over the divide above town the Professor noticed the general depression of the party, and forthwith began philosophising.

"My friends," said he, "had the feelings which explorers suffer, when fairly launched, been allowed to be present during the days of preparation, science and discovery would be in their infancy. Enthusiasm bridges the first obstacles to an undertaking, but others roll on and block the explorer's path, and the spirit which has got him into the difficulty momentarily deserts him. If properly courted, however, she returns, and meanwhile the traveler is afforded the opportunity of looking, through matter-of-fact spectacles, along his future journey. What he thought pebbles reveal themselves as hills, and what he had marked on his chart as hills develop into mountains. These he must recognize and examine with all the resolution he can summon, and he will be the more able to climb them from expecting to do so. Right here is the critical point in his journey. Numerous cross-roads branch off—some right, others left, but all with a brighter prospect down them. Perhaps on one, a wife and children stand at the door of their home, beckoning him. The garden that his own hand planted blooms in a background of flowers, while the path he has now chosen sparkles with winter snow. He knows, however, that beyond these, perhaps amid sterile mountains, are the precious diamonds he seeks.

"It is wise that, where these roads branch off—some to castles of indolence, others to comfortable homes and moderate exertion—the man should be left alone for a time and allowed to survey the rough path before him, with all the blinding glamour of enthusiasm subdued by the light of truth, and with a full knowledge of all the stumbling blocks which lie before him. If he then thumbs the edge of his hunting-knife, examines his Henry rifle, and presses forward, the metal is there, and from that time onward you may at any time learn of his whereabouts by inquiring at the temple of fame."

Sachem interrupted the Professor to remonstrate at the girding of loins being left out. He had always been used to the girding in similar discourses, and considered that loins were in much more general use than Henry rifles.

And now Shamus, from his perch on the pans, suddenly broke in: "Faith, Professor, your enthusiasm once brought me sore trouble. It got me into a haunted house, when the clock was strikin' midnight, and my legs were sore put to it to get me out fast enough. Ye see, I bet a pig with my next cousin that I would stay all night in an old house full of spirits. The master and his house-keeper had been murdered in the tenantry riots, and the boys that did the business, they swung for it soon afterward. And now, there was a regular barricadin' and attackin' going on those nights ever since. While I was lookin' at the old clock, and thinkin' of the pig I 'd drag home in the morning, I must have dramed a little. He was as likely a pig as yez ever saw, and I was listenin' proudly to his swate cries as I carried him from the sty, and feelin' full enough of enthusiasm to stay there a hundred years. Just then there was a rustlin' in front, and I opened my eyes wide, and there stood the old house-keeper leanin' against the shaky clock, with her ear to its yellow face, and lookin' straight behind me to where I could feel the master was sittin'. There was an awful light in her eyes, and I thought I heard her say—any way, I knew she was sayin' it—'Hark, Sir Donald, they 're comin', but the soldiers will be here, too, at twelve.' An' then there was a sort of shudder in the old clock and it commenced a wheezin' an bangin' away, a tryin' to get through the strokes of twelve, as it did twenty years before. But it had n't got out half, when I heard the crowd outside scrapin' against the window sill. An' then there come a report, and the room was filled with smoke, an' somethin' hit the back of my head. How I got out I do n't know, but when I come to myself I was running for dear life across the common. I have the scar of the ghost's bullet ever since. See here, yez can see it for yourselves." And taking off his cap, Shamus showed us a bald spot about the size of a silver dollar on the back of his cranium.

"And what became of the pig?" asked Mr. Colon quietly.

"Faith, an' my cousin carried him home next morning," replied Shamus, with a regretful sigh; "and lady Dobeen, bless her sowl, never forgot to tell me of that to her dying day. We were needin' the bacon them times."

Sachem, who delighted to spoil our cook's stories, declared that, to gain a pig, it was worth the cousin's while to fire an old musket through the window over a drunken Irishman inside. Still that did not excuse him for his carelessness; he should have seen that the wad flew higher.

What Dobeen's answer might have been will never be known; for, just at that moment, the attention of the entire party was suddenly directed to a dark mass of moving objects away off upon our right, a mile distant at least, and to our untrained eyes entirely unrecognizable. The Mexicans, however, pronounced them buffaloes. Whether thinking to vindicate his reputation for personal courage, or whether simply from love of excitement, is not exactly clear, but Dobeen eagerly requested permission to pursue them, and as he would, *ex officio*, be debarred the pleasure of future sport, consent was given. This was done the more readily, because we knew that Shamus, while as inexperienced in the chase as any of us, was also a wretched rider; for, although constantly boasting of the tournaments he had been engaged in, we all indorsed Sachem's opinion, that, if ever connected with such an affair at all, it must have been in holding a horse, not riding one.

It was worthy of note that every one of the party was as eager for the chase as Shamus, and yet that personage was allowed to ride off alone. Mr. Colon, it is true, essayed to join his company, but after going a hundred yards or so, suddenly changed his mind and came back. Our maiden efforts in buffalo hunting promised such modesty as to refuse a public appearance, unless together.

Our cook had been instructed by the guide to avail himself of the ravines, and after getting as near the herd as possible, then spur rapidly up to it. He went off at a gallop, his solid body flying clear of the saddle whenever the donkey's feet struck ground, and soon disappeared in a ravine which seemed to promise a winding way almost into the very midst of the herd. We watched intently for his reappearance. In such periods of suspense the minutes seem strangely long, creeping as slowly toward their allotted three-score as they do when one, at a sick-bed vigil, listens for the funeral chimes of the clock, telling when the minutes are buried in the hours.

At length, in the far away distance, we descried Shamus, disdaining further concealment, riding gallantly out of the ravine for a charge. A few moments more and game and hunter were face to face, and we held our breath, expecting to see the dark cloud dash away with our bloodthirsty cook at its skirts. "As I am alive," suddenly ejaculated Muggs, "Dobeen's coming this way, at a bloody good run, and the buffalo after him!" We could scarcely believe our eyes, but, sure enough, it was a clear case of pursuer and pursued, with the appropriate positions entirely reversed. Shamus seemed imitating that famous hunter who brought home his bear-meat alive, preceding it by only half a coat-tail. But the game before us was changing in appearance most wonderfully. It seemed bristling with un usually long horns, and as we looked the dark cloud suddenly spread out into a fan-like shape, and we all cried, simultaneously, "Indians!"

There they were, a party of our red brethren bearing rapidly down upon us in pursuit of Dobeen, whose arms and legs were playing like flails on his donkey's sides, with an appeal for speed which had evidently called into action all the reserves of that true conservative.

Our party would have sold out their interest in the plains for a bagatelle. Our whole outfit had whirled, like a weather-cock, and was pointing back to Hays. The Mexicans were already dodging in and out among their oxen, and firing their old muskets furiously, although the foe was yet a fair cannon-shot away. Shamus could not well have been in more danger from foes behind than he was from friends before; indeed, he afterward said that asking deliverance from the latter made him almost forget the former.

The horses of both Sachem and Muggs ran away, taking a straight line for the distant town. This caused a general stampede on the part of all the other horses, much to the regret of their riders, who were thus cruelly prevented from a proper display of latent prowess in rendering protection to the wagons and our cook. From the former came a steady cannonade. Squirming like eels among their oxen, the Mexicans fired from under the animals' bellies, astride the tongue, from anywhere, indeed, that furnished a barricade between the distant Indians and themselves.

OUR HORSES RUN AWAY WITH US.

It is one of the remarkable tactics of this remarkable people, in military emergencies, that when they can not put distance between them and the enemy, they must substitute *something* else. A single trooper, on an open plain, could send a small army of them scampering off, but let them get behind a barricade, and they will continue banging away with their old muskets until either the weapon bursts or ammuition gives out. It is surprising how harmless their fusillades generally are. If Mexican powder is used, it goes off like a mixture of lamp-black and nitroglycerine, with a premonitory fiz and then a fearful concussion, leaving a smell of burnt oil in the air which overcomes for a moment the natural aroma of the warriors themselves.

But while we were still being run away with by our spirited animals, another change occurred in the situation equally as unexpected as the first. The Indians had stopped running about the time that we commenced, and now stood in a dusky line something less than half a mile off, making signs to us. Shamus evidently considered it a horrible incantation for his scalp, and every time he looked backward plied with renewed fervor at his donkey's ribs. Our guide, who had stayed with the wagons and exerted himself to silence the Mexican batteries, motioned us to return, which we were finally enabled to do by virtue of steady pulling upon one rein and coming back in half circles.

By the time our cook reached us, out of breath and perspiring terribly, two savages had ridden out from their band, weaponless, and were now gesturing a wish to communicate. The Professor and our guide rode to meet them, apparently unarmed; but with characteristic exhibition of the white man's subtlety, the tail-pocket of the philosopher's coat held a pistol in reserve, and the guide, I have no doubt, was equally well provided.

CHAPTER XI.

WHITE WOLF, THE CHEYENNE CHIEF—HUNGRY INDIANS—RETURN TO HAYS—A CHEYENNE WAR PARTY—THE PIPE OF PEACE—THE COUNCIL CHAMBER—WHITE WOLF'S SPEECH, AS RENDERED BY SACHEM—THE WHITE MAN'S WIGWAM.

ABOUT midway between our party and the dusky group that stood watching us the four embassadors met. The Indians proved to be a band of Cheyennes, under White Wolf, or, as he is more frequently called, Medicine Wolf, out on the war-path against the Pawnees. The Wolf was a fine-looking man, six feet four in height, straight as an arrow, and developed like a giant. Being a chief, he possessed the regalia and warranty deed of one, consisting of a ragged military coat without any tail, and a dirty letter from some Indian agent, with a lie in it over which even a Cheyenne must have smiled, telling how White Wolf loved the whites. Perhaps he did; his namesake loves spring lamb.

Our guide was an indifferent interpreter, but had no difficulty in understanding that the Indians were hungry and wished something to eat. In all my experience from that day to this I have never found an Indian who was not hungry, except once. The exception was an old fellow who, although enough of an Indian to be habitually drunk, was so degenerate a specimen in other respects as to be somewhat dyspeptic. His stomach had repudiated, after receiving a deposit from a trader of one hundred pickled oysters, and had temporarily closed its doors. His stock of gastric juices seemed to have been well-nigh bankrupted by a fifty years' discounting of jerked buffalo. The one hundred tons of this compound which the noble warrior had dissolved would have exhausted the liquid of a tannery. Let these savages of the plains meet a white man, whenever or wherever they may, their first demand is always for meat and drink, followed not unfrequently by another for his scalp. The victim may have but a day's rations, and be a hundred miles from any station where more can be obtained, but his all is taken as greedily and remorselessly as if he commanded a commissary train.

The Professor and our guide motioned White Wolf and his companion to wait, and rode back to us for the purpose of casting up our account of ways and means. The only chance of balancing it seemed to be by sight draft on Shamus' wagon or an entry of war. We dare not refuse them and go on; they would be sure to dog our steps, and at the first convenient opportunity attack and probably murder us. Shamus, with recovered courage, stoutly protested against a raid upon his department. "To think," he expostulated, "of the swate sausage and ham bein' used to wad such painted carcasses as them divils!" The guide suggested as the best alternative that we should invite the Indians to return with us to Hays. We caught at the idea and adopted it immediately; and while the guide rode back as the bearer of our invitation, we "stood to arms," awaiting the result with silent but ill-concealed solicitude.

Should the Indians consider it an attempt to trap them, our bones might have an opportunity to rest in some neighboring ravine until the ready spades of some future geological expedition should disturb them, and we be at once reconstructed into some rare species of ancient ape or specimens of extinct salamanders. Or, if happily resurrected at a somewhat earlier period, might not some enterprising Barnum of the twentieth century place on our bones the seal of centuries, and lay them with the mummies in his show-cases? Our expedition was partly intended for diving into the past, but not quite so deep or so permanent a dive as that. What wonder that incipient ague-chills played up and down and all about our spinal column, as we reflected how completely we were dependent on the caprice of those Native Americans sitting out there, in half-naked dignity, on their tough ponies? Or that we gazed anxiously at the huge chief as he sat, silent and motionless, awaiting the approach of our guide?

Our ideas of the savage had been so thoroughly Cooperised during boyhood, that when our guide approached the Wolf, and, with a gesture to the south, invited him back to Hays, I was prepared to see the tall form straighten in the saddle, and pictured to my imagination some such specimen of untutored eloquence as this:

"Pale-face, the blood of the Cheyenne burns quick. He meets you trailing like a serpent across his war-path, seeking to steal treasures from the red man's land. He asks food, and you tell him to come into your trap and get it. Pale-faces, remove your hats; noble Cheyennes, remove their scalps!"

Nothing of this kind occurred, however. Our guide informed us that the bold savage simply fastened one button of his tailless coat, grunted out "Ugh!" in a satisfied way, and motioned his band to follow. This they did, and we were soon retracing our steps to Hays; by the guide's advice, making the savages keep a fair distance behind us.

The roofs of Hays glistened across the plains, as they say those of Damascus do in the East. We had formed a boy's romantic acquaintance with that land, where the sun burns and the simooms frolic, and once were quite enamored of its wild Bedouins of the desert. Our manhood was now experiencing the sensation of seeing a tribe fiercer than their eastern brethren, not exactly at our doors, because we had none, but following very closely at our heels.

As our strange cavalcade re-entered the town the people stopped to gaze a moment, and then came out to meet us. News flew to the fort, and some of the officers rode over. The Land Company's office was selected for a council room, the Cheyennes tying their ponies to the stage corral near. The Indians were a strange-looking crew. Sachem declared them all women, and Dobeen affirmed that they looked more like a covey of witches than warriors. With their long hair divided in the middle, and falling, sometimes in braids and again loosely, over their shoulders, and their blankets hanging around them, they did really look much like the traditional squaw who so kindly assists one in cutting his eye-teeth at Niagara Falls, with her sharp practice and cheap bead-work. Their faces were as smooth as a woman's, without the least trace of either mustache or whiskers; so that, altogether, when we essayed to pick out some females, we got completely "mixed up," and were at length forced to the conclusion that the majestic White Wolf was traveling over the plains with a copper-colored harem.

Cooper having told us that the Indian term of reproach is to be or to look like a woman, we avoided offense and the "arrows of outrageous fortune" which an Indian is so dexterous in using, and gained the information desired by addressing a direct inquiry to White Wolf, through the interpreter, whether he had any squaws along. He replied by holding up two fingers and pointing out the couple thus designated. We tried to find, first in their features and then in their clothing, some distinguishing characteristic but found it impossible; so that when they changed positions an instant afterward, I was entirely at a loss to recognize them again.

All had extremely uninviting countenances, any one of which would have sufficed to hang three ordinary men, and a common villainy made them as much alike as forty-six nutmegs. White Wolf alone differed in appearance. He was stoutly built, as well as tall and straight, with broad features, the bronze of his complexion merging almost into white, and he smiled pleasantly and readily. The others were no more able to smile than Satan himself, the expression which their faces assumed when attempting it being simply diabolical. Dobeen was so startled by one who tried that contortion on and asked for "tobac," that he retreated in disorder from the council-chamber.

White Wolf and the more important members of his band took the chairs proffered them, and sat in a circle, the Professor, Sachem, and two leading citizens of Hays being sandwiched in at proper intervals. The object of the gathering was gravely announced to be that the Indians might smoke the pipe of peace with the towns-people. As war was a chronic passion with these wild horsemen of the plains, none of them had ever been near the place in friendly mood before, and the novelty of the occasion, therefore, brought the entire population around the building. The postmaster of Hays, Mr. Hall, had once traded among the Cheyennes and, understanding their sign-language, acted as interpreter. This curious race has two distinct ways of conversing—one by mouth, in a singularly unmusical dialect, and the other by motions or signs with the hands. The latter is that most generally understood and employed by scouts and traders.

One of the Indians now took from a sack a red-clay pipe, with a ridiculously long bowl and longer shank, and inserted into it a three-foot stem, profusely ornamented with brass tacks and a tassel of painted horse hair. This was handed to White Wolf, together with a small bag of tobacco, in which the Killikinnick leaves had been previously crumbled and mixed. These were a bright red, evidently used for their fragrance, as they only weakened the tobacco without adding any particular flavor. We were struck with

THE PIPE OF PEACE—THE PROFESSOR'S DILEMMA.

the Indian mode of smoking. The chief took a few quick whiffs, emitting the fumes with a hoarse blowing like a miniature steam-engine. He then passed it, mouth-piece down so that the saliva might escape, and it commenced a slow journey around the circle. When it reached our worthy professor he found him self in a sore dilemma. No smoke had ever curled along the roof of his mouth, or made a chimney of his geological nose. For an instant the philosopher hesitated; then, reflecting that passing the pipe would be worse than choking over it, the excellent man put the stem to his mouth and gave a pull which must have filled the remotest corner of his lungs with Killikinnick. Gasping amid the stifling cloud, it poured from both mouth and nose, and called on the way at his stomach, which gave unmistakable symptoms of distress. We feared that he would be forced to forsake the council, but, with an effort worthy of the occasion and himself, he kept his seat, and opening wide his mouth, waited patiently until the fiend of smoke had withdrawn from his interior its trailing garments.

The council disappointed us. In White Wolf we had found as fine-looking an Indian as ever murdered and stole upon his native continent. His people were first in war, first to break peace, and the last to keep it, their excuse being that the white man trespassed on their hunting grounds. We had rather expected that burly form to rise from his seat, and, with flashing eyes, utter then and there a flood of aboriginal eloquence: "White man, your people live where the sun rises, ours where it sets. When did you ever come to us hungry and be fed, or clothed and go away so," and so on *ad infinitum*. Instead of all this there was a tremendous smoking and grunting, more like a farmer's fumigation of hogs than one of those pipe-of-peace councils which I had so often studied on canvas and in books. I have often regretted since that our aborigines can not read. If they could only learn from the white man's literature what they ought to be, the contrast between it and what they really are would be so violent that it might make an impression, even upon an Indian.

For a happy mingling of lies and truth our "big talk" could hardly be excelled. A reporter could have taken down the proceedings somewhat as follows:

Scene—Six Indians and as many white men in a ring. Postmaster Hall in the center, acting as interpreter.

Indian—"Cheyenne love white man much (lie). Forty-six warriors all hungry (truth). Us good Indians" (lie). And so on, alternately.

Pale Brother—"White man love Cheyenne. Got lots of food, but no whisky" (the latter a lie which almost choked the speaker).

It would not interest the reader to know all the repetitions or nonsense uttered, and we spare him the infliction of even attempting to tell him. The Indians had for their object food, and they got it. The whites had for their object permanent peace, and did not get it.

In due time the council broke up, and in an incredibly short time thereafter many of the Indians were reel-

WHITE WOLF AT HOME.

"The red man is noble, big injun is me."

ing drunk. That White Wolf did not become equally so was owing altogether to his being a man of iron constitution. Any thing but metal, it seemed to me, must have been burnt out by the fiery draughts which we saw the noble chief take down. A tin cupful of "whisk," such as would have made the cork in a bottle tight, was tossed off without a wink.

Sachem, who took notes, rendered White Wolf's speech at the council in verse, as follows:

White brother, have pity; the White Wolf is poor,
The skin of his belly is shrunk to his back;
A gallon of whisky is good for a cure,
If followed by plenty of "bacon and tack."

The red man is noble, big Injun is me:
Like berries all crimson and ready to pick,
The scalps on my pole are a heap good to see—
Good medicine they when poor Injun is sick.

The red man is truth, and the white one is lies;
The first suffers wrong at hand of the other;
The way they skin us is good for sore eyes,
The way we skin them astonishing, rather.

They rob us of guns and offer us plows,
And tell us to farm it, to go into corn;
We 're good to raise hair, and good to raise rows,
And good to raise essence of corn—in a horn.

Go back to your cities and leave us our home,
Or off with your scalp and that remnant of shirt;
Go, let the poor Injun in happiness roam,
And live on his buffalo, puppies, and dirt.

Two or three of the Indians mounted their ponies and took a race through the streets. The animals were thin, despondent brutes, but as wiry as if their hides were stuffed, like patent mattresses, full of springs. The Indians, as is their universal custom, mounted from the right side, instead of the left as we do. At the lower end of the street they got as nearly in line as their inebriated condition would permit, and when the word was given set off toward us with frightful shouts, which made the ponies scamper like so many frightened cats.

The animal which came out ahead had no rider to claim the honors, that blanketed jockey having fallen off midway. He was now sitting on his hams, looking the wrong way down the track, and evidently adding up the "book" which he had made for the race. As he soon arose, with a dissatisfied grunt, we thought his figures probably read about as follows:

Given—A gallon of Hays whisky in the saddle, and a race-horse under it. Endeavor to divide the latter by a rawhide whip, and the result is a soreheaded Indian, who stands forfeit to his peers for "the drinks."

As we wandered back to the council-chamber, the scene there had changed somewhat. White Wolf had been transformed into a cavalry colonel, and was strutting around with two gilt eagles on his broad shoulders, looking fully as important as many a real colonel whom we have caught in his pin feathers and, withal, much more of the hero. Our warrior had seen some of the officers from the fort strolling around, and straightway fell to coveting his neighbors' straps, which observing, Sachem at once purchased from a store the emblems of power and pinned them upon him. He whispered to us that when White Wolf took his first step as a colonel, it had been accompanied by a snort of pain, the unlucky slipping of a pin having evidently conveyed to the chief the idea that one of the eagles had grasped his shoulder in its talons.

The chief modestly requested similar honors for his "papoose," and that individual was treated to the straps of a captain. A different application of strap, it occurred to me, would have seemed more proper upon the six feet of unpromising humanity which appeared above the "papoose's" moccasins.

It had been a matter of surprise to us how the Indians could make such inferior looking stock as theirs capable of such speed and extraordinary journeys; but it ceased to excite our wonder after an examination of their whips. These ingenious instruments of torture have handles, which in form and size resemble a policeman's club. To one end are attached some thongs of thick leather, half a yard in length, and to the other a loop of the same material, just large enough to go over the hand and bind slightly on the wrist. Dangling from the latter, the handle can be instantly grasped, and the body of thongs brought down on the pony's skin, with a crack like a flail on the sheaves, and the result is what Sachem called an astonishing "shelling out" of speed.

We explained to White Wolf that Tammany

Sachem was one of many great chiefs who had a mighty wigwam in the big city of the pale-faces, far away toward the rising sun; that they were all good men, and never lied like the chiefs of the Cheyennes, or took any thing belonging to others; and that their women, instead of carrying heavy burdens, spent all their time in distributing the money and goods of the big wigwam to the needy.

White Wolf signified, through the interpreter, that such a wigwam was too good for earth, and ought to be pitched on the happy hunting grounds as soon as possible.

Sachem thought the savage meant to be sarcastic.

CHAPTER XII.

ARMS OF A WAR PARTY—A DONKEY PRESENT—EATING POWERS OF THE NOMADS—SATANTA, HIS CRIMES AND PUNISHMENT—RUNNING OFF WITH A GOVERNMENT HERD—DAUB, OUR ARTIST—ANTELOPE CHASE BY A GREYHOUND.

AT our request White Wolf and two of his braves gave us a display of their skill—or rather, their strength—in the use of their bows, shooting their arrows at a stake sixty yards off. The efforts were what would be called good "line shots," although missing the slender stick. We then essayed a trial with the chief's bow, which was an exceedingly stout hickory wrapped in sinew, but we found that more practiced strength than ours was required even to bend it. Some amusement was created when the first of our party took up the bow, by the haste with which a small and unusually ugly Indian retreated from the foreground as if fearing that an arrow might be accidentally sent through his blanket.

Among the stock which the savages had brought with them was a long-eared, diminutive brute, scarcely higher than a table, and apparently forming the connecting link between a jackass rabbit and a donkey. This animal White Wolf seemed extremely anxious to present to the Professor, but it was politely declined, by the advice of the interpreter, who explained to us that a return gift of the donkey's weight in sugar and coffee would be expected. Notwithstanding the stringency of the law forbidding the sale of whisky and ammunitions to the Indians, the savages found little difficulty in filling themselves with fire-water, and also got a little powder. White Wolf went off with his pocket full of cartridges in exchange for some Indian commodities, but the cunning pale face rendered them of little value by selecting ammunition a size too small for the gun.

The eating powers of these nomads are marvelous. We saw the chief, inside of two hours, devour three hearty dinners, one of which was gotten up from our own larder and was both good and plentiful. As he did full justice to every invitation to eat and drink, we concluded that he would continue to accept during the whole afternoon, if the opportunity were only offered him. What a capital minister to England was here wasting his gastric juices on the desert air! If Great Britain should continue her hesitation to digest our Alabama claims, the wolf at their door would digest enough roast beef to bring them to terms or starvation. Sugar, coffee, spices, pickles, sardines, ham, and many another luxury of civilization, were alike welcome at the capacious portal of the untutored savage. Dobeen discovered him eating a can of our condensed milk under the impression that it was a sweet porridge.

Their entertainment at the town being concluded, the Indians were conducted over to the fort and some rations given them. They manifested an especial fondness for sugar, but took any thing they could get, their ponies proving capable of carrying an unlimited number of sacks. It seemed as difficult to overload these animals as it is a Broadway omnibus; and their riders, perhaps in order to avoid being top heavy, took freight for the inside whenever opportunity offered. As they came back through the town, we all turned out to see them off. The band promised us peace, notwithstanding which it was no small satisfaction to discover that they were poorly armed. Bows and arrows were the only weapons which all possessed, and while a few had revolvers, the chief alone sported a rifle, a rusty-looking old breech-loader.

As our late cavalry escort rode off, their attitudes plainly bespoke that they had been raiding upon more than the flesh-pots of Egypt. Sons of the sandy-complexioned desert, we saw several of them kiss their mother before they got out of sight. The most serious question with us now was whether or not these red gormandizers had been uttering peace notes not properly indorsed by their hearts. The trouble is that when one discovers a circulation of this kind, his own ceases about the same instant, and his bones become a fixed investment in the fertile soil of the plains.

One of the officers of the fort told us an amusing instance of the impudent treachery of which the western Indians of to-day are sometimes guilty. A year or two before, when Hancock commanded the Department and was encamped near Fort Dodge, on the Arkansas, Satanta and his band of Kiowas came in. This chief has always been known as very hostile to the whites, usually being the first of his tribe to commence hostilities. He was the very embodi ment of treachery, ferocity, and bravado. Phrenologically considered, his head must have been a cranial marvel, and the bumps on it mapping out the kingdom of evil a sort of Rocky Mountain chain towering over the more peaceful valleys around. Viewed from the towering peaks of combativeness and acquisitiveness the territory of his past would reveal to the phrenologist an untold number of government mules, fenced in by sutler's stores, while bending over the bloody trail leading back almost to his bark cradle, would be the shades of many mothers and wives, searching among the wrecks of emigrant trains for flesh of their flesh and bone of their bone.

Satanta was long a name on the plains to hate and abhor. He was an abject beggar in the pale faces' camp and a demon on their trail. On the occasion in question he came to Gen. Hancock with protestations of friendship, and, although these were not believed, he was treated precisely as if they had been. To gratify his love of finery an old military coat with general's stars, said to be one that Hancock himself had cast off, was presented him. By some means he also acquired a bugle, and the garrison were greatly amused for the remainder of the day by seeing Satanta galloping back and forth before his band, blowing his bugle and parading his coat, the warriors all cheering the old cut-throat and proud as himself of the display. The way he handled that bugle, however, before the next morning was by no means so amusing.

Some time before dawn the sleepy garrison were aroused by the thunders of a stock stampede, and out of the darkness came the clatter of hoofs, as Satanta and his band departed for the south with a goodly herd of government mules and horses. Pursuit was commenced at once, with the hope of cutting them off before they could get the stock across the Arkansas, then somewhat swollen. Just as the troops reached the bank of that stream, a major-general's uniform was seen going out of the water upon the other side. Notwithstanding its high rank fire was instantly opened upon it, but ineffectually. The savage turned a moment, blew a shrill, defiant blast upon his bugle, and galloped off in safety. Too much promotion made him mad. As a simple chief, he might have stolen some straggling teams; as a major-general, he appropriated a whole herd.

During the next eighteen months, Satanta had several encounters with the troops, generally wearing the major-general's coat and blowing his bugle. His last exploit, which brought the long hesitating sword of justice upon his head, is too fresh and too painful to be soon forgotten. A few months ago the savage chief was living with his people on a reserve in the Indian Territory and being fed by the government. Gathering a few of his warriors he stole forth, and, crossing the Texas border, surprised a wagon train, murdered the teamsters, and drove off the mules. Fortunately, Gen. Sherman, in his examination of frontier posts, happened to be near the scene of murder, and at once ordered troops in pursuit. They were still trailing the marauders when Satanta returned to the reservation at Fort Sill, and with bold effrontery begotten of long immunity, actually boasted of the crime before the Quaker agent. "I did it," said he, "and if any other chief says it was him, tell him he lies. I am the man." Gen. Sherman had just arrived, and when Satanta, with a number of minor chiefs who were with him on the raid, came into the fort to trade and visit, they were seized and bound, and started for Texas under a strong guard, to be tried by the authorities there. On the way one of the Indians in some manner loosened his bands, and seizing the musket of the guard nearest him, shot the soldier in the shoulder, but before he could do further harm the other guards fired, and the savage rolled from the wagon down upon the plain, apparently dead. The body was afterward found close by the road-side in a position which showed that after falling the savage had enough of vitality left to enable him to crawl with bloody hands for several yards. Finding the life-tide ebbing fast, he had then placed his body in position toward the rising sun, composed his arms by his side and, with Indian stoicism, yielded up his breath. The remainder of the party, including Satanta, were brought safely to Texas, tried, and sentenced to be hanged.

Our adventure with White Wolf and his band obliged us, of course, to pass another night in Hays. We spent a most pleasant hour during the evening in the office of Dr. John Moore, an old resident of

Plattsburg, N. Y., who assisted us materially in selecting medical stores, and who by his genial disposition endeared himself to our entire party, so that when we heard of his sad fate soon afterward, it seemed as if death had crouched by our own camp-fire. Should the Indians become troublesome, there was some talk at the fort, he now informed us, of organizing a company for operations against them, composed of buffalo hunters and scouts under the lead of regular officers, and in this case it was his purpose to accompany it in the capacity of a surgeon. As good guns were difficult to obtain there, and we had some extra weapons, one of our party loaned the doctor an improved Henry rifle and holster revolvers. Before we again heard of him, he had crossed that shadowy line which winds between the tombs and habitations of men, and his name was added to the drearily long list which bears for its heading—"Killed by Indians."

Commencing with those first entries after the Mayflower introduced our fathers to savage audience, and chiseling separately each name on a marble milestone, the white witnesses would girdle the earth.

Sunrise next morning saw us again moving northward, fully determined that no body of Indians, unless comprising the whole Cheyenne nation, should force us back again. We had met the red man on his native heath and familiarity had bred contempt. All were in excellent spirits and felt the braver, perhaps, because our late visitors had assured us that their tribe was on the war-path against the Pawnees, and meant only peace with the whites.

Our party left Hays the second time with quite an acquisition. On the eve of starting we had been approached by an artist, who begged permission to accompany us. We assented on the instant. An artist was, of all others, the thing we needed. How interesting it would be to have the thrilling incidents of the coming months sketched by our artist on the spot. "Daub" was a fine-looking fellow, with peaked hat, peaked beard, and peaked mustache; in short, was of the genuine artist cut, of the kind that are always sitting around on the stones in romantic places and getting married to heiresses.

During the day we saw many varieties of the cactus, some of them very beautiful. As we had no regular botanist with our expedition, Mr. Colon developed a taste in that direction, and secured and deposited several fine specimens which were carefully laid away in Shamus' wagon. It was not long before that excellent Irishman gave a prolonged howl, the cause of which he did not vouchsafe to tell us, but as we saw him cautiously rubbing his pantaloons we surmised that he had rolled or sat down upon a choice variety. The remainder of the plants he must, with still greater caution, have dropped overboard, as none could subsequently be found for boxing. If the truth must be said, I was not at all sorry for it. I had lent a hand in obtaining an unusually large cactus, but the loan was returned in such damaged condition that I lost all interest at once. The minute needles which nature has scattered over these plants will pierce a glove readily, and burrow in the flesh like trichina. The cactus may be set down as Dame Nature's pin-cushions.

Endless prairie-dog villages covered the country, and occasionally cayotes, about the size of setters, with brushy, fox-like tails, started out of ravines and ran off with a hang-dog sort of look, stopping occasionally to see if they were being pursued. Our guide ran one of these down with his horse and it was almost with sympathy that we watched the tired wolf, when he found running useless, dodging between the horse's legs, rendering the rider's aim false. It was finally dispatched by a greyhound. The latter deserved his name only from courtesy of species, as his color was inky black. He belonged to one of our hostlers, who got him from a Mexican train-master, and was a wonderful fighter. I saw him afterward in combats with not only the cayote, but the large timber wolf, and in every instance he came off the victor. On one occasion, I remember, he whipped the combined curs of a railroad tie camp, making every antagonist take to his heels. Very nearly as high as a table, with powerful chest and immense spring, the hound's movements were like flashes of light. He danced round and over his foe, his fangs clicking like a steel trap, first on one side and now on the other, and again, ere his enemy had closed its jaws on the shadow in front, he was at the rear I have seen a gray wolf bleeding and helpless, and the hound untouched, after a half hour's combat.

On the north fork of Big Creek we frightened a dozen antelopes out of the brakes, and had a fine opportunity of witnessing a chase by the hound which alone was worth a journey to the plains to see. I remember having been very much interested, when a boy, in reading accounts of gazelle hunting in the Orient, where hawks and dogs are both used. The former pounce down from the air on the fleet-footed victim's head, compelling it to stop every few moments to shake its unwelcome passenger off, and the dogs are thus enabled to overtake it. This always seemed to me a cowardly sort of sport. The harmless victim of the chase, who can not touch the earth without its turning tell-tale to the keen-scented pursuer, should not be robbed of his only refuge, speed, or the pursuit becomes butchery.

The American antelope upon our plains is what the gazelle is upon those of Africa. Timid and fleet, it often detects and avoids danger to which its powerful neighbor, the buffalo, falls a victim. The group which we had frightened bounded away with an elasticity as if nature had furnished them hoofs and joints of rubber. There was no apparent effort in their motion, and we imagined larger powers in reserve than really existed. As the greyhound slowly gained upon them, we noticed this, and the Professor thereupon delivered what Sachem aptly styled a running discourse.

"Gentlemen, poetry of motion, perhaps by poetical license, gives exaggerated ideas of force. A smooth-running engine, though taxed to its utmost capacity, seems capable of accomplishing more, while its wheezing neighbor, groaning and straining as if on the verge of dissolution, has abundant powers in reserve. Some Hercules may lift a weight on which a straw more would seem to him large enough to sustain the traditional drowning man. The feat marks itself by a life-long backache, but, if he has performed it gracefully, he bears with it a reputation for a fabulous reserve of power, the exhibition seeming but the safety valve to his supposed giant forces struggling for expression."

Our learned friend seldom found us less attentive than then. All the wagons were stopped, and from every elevation upon them we looked out over the solitudes at the race going on before us. Pursuer and pursued were pitting against each other the same quality—speed. There was no lying in ambush or taking unawares. The fleetest-footed of game was flying before the swiftest of dogs. There could be no trailing, as these hounds run only by sight. What a straining of muscles! The low ridge barely lifting the animals against the horizon, their legs, from rapidity of motion, were invisible, and the bodies, for a short space, seemed floating in air. It was one short, black line, running rapidly into twelve gray ones, these latter resolving occasionally into as many balls of white cotton, when the puffy, rabbit-like tails of the antelopes were turned toward us. Two of the best mounted horsemen from our party had started with the chase, but seemed scarcely moving, so rapidly were they left behind.

Twice we thought the hound had closed, but instantly succeeding views showed daylight still between, although the narrow strip was being blotted out with the same regular certainty with which the dark slide of the magic lantern seizes the figures on the wall. Down into a ravine, and out of sight they passed, and we were fearing the *finale* would be hidden, when they came into view on the opposite side and pressed up the bank. The bounds of the hound were magnificent, and we all gave a cry of admiration, as with a splendid effort he launched himself like a black ball upon the herd. In an instant after we saw him hurled back and taking a very unvictor-like roll down the hill. He quickly recovered, however, and fastened on an antelope which seemed lagging behind. His first selection, the leader of the herd, had proved an unfortunate one, and he bore a bruise for some time where the buck had struck him with his horns.

The second seizure turned out to be a doe, and was quite dead when we reached it. The victor was lying along side, looking very much as if one antelope hunt a day was sufficient for even a greyhound. We noticed that the hair was rubbed off from the doe's sides by its struggles, and on passing our hands over the neck found that its coarse coat parted from the skin at a slight touch. This peculiarity in the antelope is very marked. In a subsequent hunt I once saw a wounded buck plunge forward, roll along the ground for a few feet, and then run off with the bare skin along his entire side showing just where he had struck the earth.

One of our party produced a knife, and the animal was bled and the entrails taken out. We seemed destined to have a mishap with every adventure, and had already learned to expect such sequences, the only question being whose turn should come next.

This time it proved to be Semi-Colon's. We were a mile from the wagons, and Semi's horse, being considered the most thoroughly broken, was nominated to bear the game to them. To this proceeding Cynocephalus seemed in nowise indisposed, quietly submitting to the management of one of the hostlers and our guide, as they lashed the antelope across his back, securing it to the rear of the large Texas saddle with the powerful straps which always hang there for purposes of this kind. This accomplished, Semi climbed into the saddle, gave a click and a kick, and set his steed in motion. That eccentric assemblage of bones made one spasmodic step forward, which brought the bloody, hairy carcass with a swing against his loins.

What a change that touch produced! Those wasted nostrils emitted a terrific snort, the stiff stump-tail jerked upward like the lever of a locomotive, and with a dart Cynocephalus was off across the plains. He probably imagined that some beast of prey had coveted his spare-ribs, and was whetting its teeth on the vantage-ground of his backbone. Occasionally the frightened animal would slack up and indulge in a fit of kicking, looking back meanwhile with terror at the object fastened upon his hide, then plunge frantically forward again. The antelope stuck to the saddle for some time, but not so Semi-Colon. The first of these irregular proceedings caused that young man, as Sachem expressed it, "to get off upon his head." Cynocephalus finally burst his saddle-girths, and we were obliged to furnish other transportation for our game.

Let me say, *en passant*, that I am trying to chronicle minutely the events which befel our half-scientific, half-sporting, and somewhat incongruous party on its trip through Buffalo Land; and, although my readers may think us particularly unfortunate, we really suffered no more than amateurs usually do. My object is to set up guide boards at the dangerous places, that other travelers may avoid the pitfalls and the perils into which we fell. And to every amateur hunter we beg to offer this advice: Never tie dead game upon a strange horse unless you owe the rider a grudge.

"Young men," said the Doctor, from his saddle, "you have seen a beautiful illustration in the theory of development. The hound and the antelope may have been originally an oyster and a worm. From their first slow motion, when one only opened its jaws to seize the other, they have progressed until the speed of to-day results. Should the hound ever become wild, and pursuit and flight change to an every-day matter instead of a holiday-sport, development would still continue. A giraffe-like antelope, with the speed of the wind, would fly before a hound the size of a stag." The Doctor's "clinic," as Sachem called it, was suddenly cut short at this point by a struggle for mastery between himself and the human spirit concealed in his horse.

"How much," exclaimed the Professor, when Pythagoras had at length come off triumphant, and we again moved forward—"How much the race that we have witnessed is like that we all run. Powerful and eager as the greyhound, man sees flying before him, on the plain of life, an object which he thirsts to grasp. Taxing every muscle in pursuit, panting after it over the smooth country below the 40th mile-post, he crosses there the ravine where rheumatism and straggling gray hairs lurk, and with these clinging to him, starts up the hill of later life. Half-way to its summit, on which the three-score stone marking the down-hill grade looks uncomfortably like that over a tomb, he seizes the object of pursuit only to be flung back by it bruised. If of the proper metal, he falls but to rise again, and should the first wish be out of reach, fastens on one of its companions. There is where blood tells. If the least taint of cur is in it the first blow sends its recipient yelling to his kennel, there to whine for the remainder of life over bruised ribs."

Muggs thought a single toss was sufficient, and retreat then only prudence. If the bones on one side were broken, he saw no reason to expose the other. Dying successful was only procuring meat for others to enjoy.

The Professor was developing a remarkable talent for finding not only the stones of the past written all over with a wonderful and translatable history, but also the moral connected with each incident of our journey. Had any of us broken our necks he would doubtless have improved the occasion to draw a comparison and have made it the text of a philosophic disquisition

CHAPTER XIII.

CHARACTER OF THE PLAINS—BUFFALO BILL AND HIS HORSE BRIGHAM—THE GUIDE AND SCOUT OF ROMANCE—CAYOTE VERSUS JACKASS-RABBIT—A LAWYER-LIKE RESCUE—OUR CAMP ON SILVER CREEK—UNCLE SAM'S BUFFALO HERDS—TURKEY SHOOTING—OUR FIRST MEAL ON THE PLAINS—A GAME SUPPER.

OUR trail was taking us west of north, and we expected to reach the Saline about dusk and there encamp. The same strange evenness of country surrounded us. Over its surface, smooth and firm as a race track, we could drive a wagon or gallop a horse in any direction. Even the Bedouin has no such field for cavalry practice—his footing being shifting sand, while ours was the compact buffalo grass, so short that its existence at all could scarcely have been detected a few yards away. Sachem said he could think of no such cavalry field except that of his boyhood, when he slipped into the parlor and pranced his rocking-horse over the soft carpet; with which memory, he added, was coupled another, to the effect that while thus skirmishing on dangerous ground, his cavalry was attacked from the rear by heavy infantry and badly cut up.

Numerous buffalo trails crossed our path, running invariably north and south. This is caused by the animals feeding from one stream to another, the water courses following the dip of the country's surface from west to east. Wallows were also very numerous, and we noticed as a peculiarity of these, as well as the paths, that the grass killed by treading and rolling does not renew itself when the spots are abandoned. More than once on the Grand Prairie of Illinois I have seen these wallows, made before the knowledge of the white man, still remaining destitute of grass.

An old bull who has been rolling when the wallow is muddy, is an interesting object. The clay plastered over and tangled in his shaggy coat bakes in the sun very nearly white; and this it was, probably, that gave rise to the early traditions of white buffalo.

Wherever on our route the rock cropped out along creeks or in ravines, it was the white magnesia limestone, and so soft as to be easily cut. Further west alternate pink and white veins occur, giving the stone a very beautiful appearance. We frequently found on the rocks and in the ravines deposits of very perfect shells, apparently those of oysters. Sachem suggested that they marked the location of pre-historic restaurants—the Delmonicos of the olden time, say fifty thousand years before the Pharaohs were born. He thought it possible that some future quarryman might blast out an oyster-knife and money pot of quaint coins.

Muggs thought this patch of our continent resembled Australia—"Not that it is as rich, you know, but there's so much of it." He even became enthusiastic enough to affirm that the land might be made profitable, "if some Hinglish sheep and 'eifers were put on it, you see."

The Professor assured us that the country around was equal to the plains of Lombardy in point of fertility, and as the soil was of great depth, and rich in the proper mineral properties, it would undoubtedly become before 1890 the great wheat-producing region of the world.

Our party fell into silence again, and, having nothing else to interest me at the moment, I resumed my study, which this episode had interrupted, of Buffalo Bill, our guide. Athletic and shrewd, he rode ahead of us with sinews of iron and eye ever on the alert, clad in a suit of buckskin. His mount was a tough roan pony which he had named Brigham and of which he seemed very fond. Nevertheless, this fondness did not prevent hard riding, and when I last saw Brigham, several months afterward, he was a very sorry-looking animal, insomuch that I concluded not to have his photograph taken as that of a model steed for Buffalo Land, as I once contemplated doing.

It was extremely fortunate for us that we had secured Cody as guide. The whole western country bordering on the plains, as we afterward learned, from sorry experience, is infested with numberless charlatans, blazing with all sorts of hunting and fighting titles, and ready at the rustle of greenbacks to act as guides through a land they know nothing about. These reprobates delight in telling thrilling tales of their escapes from Indians, and are constantly chilling the blood of their shivering party by pointing out spots where imaginary murders took place. Without compasses they would be as hopelessly lost as needleless mariners. I have my doubts if one-third of these terribly named bullies could tell, on a pinch, where the north star is. Unless they chanced to strike one of the Pacific lines which stretch across the plains, a party, under their guidance, wishing to go west would be equally liable to get among the Northern Siouxs or the Ku-Klux of Arkansas.

A thousand miles east Young America's cherished ideal of the frontier scout and guide is an eagle-eyed giant, with a horse which obeys his whistle, and breaks the neck of any Indian trying to steal him. In addition to its wonderful master, the back of this model steed is usually occupied by a rescued maiden. At risk of infringing on the copyrights of thirty-six thousand of the latest Indian stories, we have obtained from an artist on the spot an illustration of the last heroine brought in and her rescuer, the rare old plainsman.*

Cody had all the frontiersman's fondness for practical jokes, and delighted in designating Mr. Colon as "Mr. Boston," as if accidentally confounding the residence with the name. In one instance, with a cry of "Come, Mr. Boston, here's a specimen!" he enticed the philanthropist into the eager pursuit of a beautiful little animal through some rank bottom grass, and brought the good man back in such a condition that we unanimously insisted on his traveling to leeward for the rest of the day.

While we thus journeyed, and, in traditional traveler's style, mused and pondered, Shamus came running back to say that we were wanted in front. "Such a goin' on in the ravine beyant as bates a witch's dance all holly!" We saw that the forward wagons had halted and the men were peering

* See illustration on page 137.

cautiously over the edge of the highland into the valley of Silver Creek, which stream wound along below, entirely out of sight until one came directly upon it. In this lonely land, the pages of whose history Time had so often turned with bloody fingers, an event slight as even this was startling. That hollow in the plain before us seemed to yawn, as if awaking in sleepy horrors, and we noticed a general tightening of reins and rattling of spurs. This maneuver was executed to prevent our horses running away again and thus rendering us incapable of supporting our advanced guard. If savages were around, our provisions must be protected, and we at once dismounted and scattered among the teams in such a way as to offer the most successful defense.

Our fears were groundless. In a few moments Cody came galloping back on Brigham, and said briefly that we should lose a fine lesson in natural history unless we hurried to the front. Truth compels me to say that we did not hanker after a close acquaintance with Lo on the rampage; yet we did earnestly desire to improve every opportunity of studying the other inhabitants of the plains, and a few moments accordingly found our whole party peering over the edge of the bluff into the valley below.

There, on a patch of bottom grass, half a dozen elk were feeding; a short distance away, a small herd of wild horses drank from the brook; while in a ravine immediately in front of us, three cayotes were attempting to capture a jackass-rabbit. What a wealth of animal life this valley had opened to us. From our own level the table-lands stretched away in all

BUREAU OF ILLUSTRATION

THE WILD DENIZENS OF THE PLAINS.

directions until striking its grassy waves against the horizon, with not a shrub, tree, or beast to relieve the clearly-cut outlines. Casting our eyes upward, the bright blue sky, clear of every vestige of clouds, arched down until resting on our prairie floor, and not even a bird soared in the air to charm the profound space with the eloquence of life. Casting our eyes downward, the earth was all astir with the activity of its brute creation.

Before we could make any effort at capture, the elk and horses winded us and fled away toward the opposite ridges, where stalking them would have been exceedingly difficult, if not impossible. Leading the mustangs was a large black stallion, which kept its position by pacing while the others ran. Buffalo Bill said this was an escaped American horse which had fled to solitude with the rider's blood upon his saddle. We noted the statement as one for future elucidation at our camp-fire. The rabbit chase in the ravine continued, and we watched it unseen for several minutes. The wolves were endeavoring to surround their victim, and cut in ahead of it whenever he attempted to get out of the ravine. Although such odds were against him, the rabbit had thus far succeeded by superior speed and quick dodging in evading his enemies; but escape was hopeless, as he was hemmed in and becoming exhausted. These tireless wolves, cowardly creatures though they are, might worry to death an elephant. A few shots terminated this scene, driving off the wolves, but killing the rabbit for whose protection they were fired. The Professor remarked that this was like a lawyer's rescue. He sometimes frightens away the persecutors, but the charges generally kill the client.

For the benefit of those of my readers who have never seen a member of that unfortunate rabbit family which has been christened by such a humiliating given name, I would state that the species is remarkable for its very long ears, and very long legs. If the reader, being a married man, desires a pictorial representation of this animal, let him draw a donkey a foot high on the wall, and if his wife does not interrupt by drawing a broomstick, he may be satisfied that his work is well done, and a life-size jackass-rabbit will stand out before him.

A mile from the scene of this adventure Silver Creek joined the Saline, and at the junction it was determined to make our camp. We descended among heavy "brakes," staying our loaded wagons with ropes from behind. Immense quarries of the soft, white limestone rose from the valley's bed to the level of the plains above, and the rains of centuries had fashioned out pillars and arches, giving them the appearance of ancient ruins staring down upon us. Mr. Colon picked up a fine moss agate and the Professor a Kansas diamond. Under the surface of the former were several figures of bushes and trees, outlined as distinctly as the images one sees blown into glass. The diamond was as large as a hazel nut and as clear as a drop of pure water, so that, notwithstanding its size, ordinary print could be easily read through it. Had it possessed a hardness corresponding with its beauty, the Professor could have enriched with it half a dozen scientific institutions. Such stones now command a fair market value among travelers, and are generally mounted in rich settings as souvenirs of their trips.

A picturesque group of some half-dozen oaks offered a good camping spot, and around it the wagons were placed for the night in a half-circle, the ends of the crescent resting each side of us upon the creek. The rule of the plains is, "In time of peace prepare for war."

Northward from us, and distant perhaps fifty yards, rippled the clear waters of the Saline, which was then at a low stage. High above it was the table-land of the plains, and the edge of this, as far as we could trace it, was dotted with the dark forms of countless buffalo. So distant as to appear diminutive, their moving seemed like crawling, and the back-ground of light grass gave them much the appearance of bees upon a board. They were crowding up to the very edge of the valley of the Saline, from whence, as we were told, they extended back to the Solomon, thence to the Republican, and at intervals all the way northward to the remote regions of the Upper Missouri.

Could the venerable Uncle Samuel go up in a balloon and take a thousand miles' view of his western stock region, he would perceive that his goodly herds of bison, some millions in number, feeding between the snows of the North and the flowers of the South, were waxing fat and multiplying. This latter fact might somewhat surprise him, when he discovered around his herd a steady line of fire and heard its continual snapping. The unsophisticated old gentleman would see train after train of railroad cars rustling over the plains, every window smoking with the bombardment like the port-holes of a man-of-war. He would see Upper Missouri steamers often paddling in a river black with the crossing herds, and pouring wanton showers of bullets into their shaggy backs. To the south Indians on horseback, to the north Indians on snow shoes, would meet his astonished gaze, and around the outskirts of the vast range his white children on a variety of conveyances, and all, savage and civilized alike, thirsting for buffalo blood. That the buffalo, in spite of all this, does apparently continue to increase, shows that the old and rheumatic ones, the veteran bulls which in bands and singly circle around the inner herds of cows and calves, are the ones that most commonly fall the easy victims to the hunters. Their day has passed, and powder and ball but give the wolves their bones to pick a little earlier.

Such were the thoughts that revolved in my mind while sitting upon one of the wagons, and dividing my attention between the tent pitching going on under the trees and the shaggy thousands which, feeding against the horizon, seemed to grow larger as the sun went down behind them and they stood out in deepening relief in the long autumn twilight. These solitudes made me think of Du Chaillu on the African deserts when night set in, and I wondered if the brute denizens there could be more interesting than those which surrounded us. Had a lion roared, I doubt whether it would have struck me as unnatural, although it might have induced a speedy change of base. It begets a peculiar feeling in one's mind, I thought, when the lower brutes surround him and his fellow-creature alone is absent. Animal organizations are every-where, blood throbbing and limbs moving, and yet the world is as solitary to him as if the planet had been sent whirling into space and no living being upon it except himself. A handkerchief, a hat, any thing which his brother man may have worn, yields more of companionship than all the life around him.

And now, through the trees, we saw several of our men running with their weapons in hand, and immediately afterward heard the rapid reports of their revolvers and rifles from the creek just below, followed by the fluttering, noisy exit of turkeys from among the trees. Some flew away, but most of them were running, and, in their fright, passed directly among the wagons. One old gobbler, with a fine glossy tuft hanging at his breast, had a hard time of it in running the gauntlet of our camp-followers, narrrowly escaping death by a frying pan hurled from the vigorous grasp of Shamus.

This class of our game birds is noted the continent over for its wildness and cunning, these qualities furnishing old hunters with material for numberless yarns, as they gather around the camp-fires and weave their fancies into connected sequence. Thus it has become a matter of veritable history that knowing gobblers sometimes examine the tracks that hunters have left to see which way they are going.

On Silver Creek the turkeys were very tame, and before it became too dark for shooting our party had killed twelve. Muggs and Sachem had combined their forces and devoted their joint attention to one of them sitting stupidly on a limb, where it received a bombardment of five minutes' duration before coming down. Our Briton explained that "the bird was unable to fly away, you see, because I 'it 'im at my first shot." To this statement Sachem stoutly demurred upon two grounds: First, that Muggs' gun had gone off prematurely, the time in question, and barely missed one of his English shoes; and, second, that the turkey showed but one bullet mark, and that wound was necessarily fatal, as it had carried away most of the head! A compromise was finally effected, and we were much edified by seeing the two coming into camp with the bird between them, sharing mutually its honors.

Great numbers of turkeys seemed to inhabit the creek, all along which we heard them, at dark, flying up to their roosts. This induced a number of our party to visit a large oak scarcely a hundred yards from camp, which one of our men had marked as a favorite resort. Proceeding with the utmost caution, under the dim shadows of approaching nigbt, we presently stood beneath the roost. Clearly defined between us and the sky were the limbs, and clustering thickly over them, like apples left in fall upon a leafless tree, we could descry large black balls, indicating to our hunger-stimulated imaginations as many prospective turkey roasts. For this special occasion our only two shot guns had been brought forth from the cases, the remainder of the party being furnished with Spencer and Henry rifles.

We had been instructed each to select our bird, and fire at the word to be given by the guide. How loud and sharp the clicking of the locks sounded, in the stillness of that jungle on the plains, as six barrels pointed upward, but their aim made all unsteady by the thumping of as many palpitating hearts. Then, in a low tone, came the words—and they seemed hoarsely loud in the painful silence around us—"Ready! Take careful aim!" "Hold!" cried the Professor, in a sudden outburst of enthusiasm; "Gentlemen, you see above us thirty fine specimens of that noblest of all American birds, the turkey. Wisely has it been said that, instead of the eagle, the turkey should have been our National"—"Fire!" cried the guide, in an agony, as the Professor, having dropped his gun, was rising to his feet, and the turkeys, alarmed by his eloquence, were preparing for flight.

And fire we did. A half dozen tongues of flame shot upward, and the roar of our unmasked battery reverberated over the solitude. The rustling and fluttering among the tree tops was terrific, and showers of twigs and bark rained down upon us. Every one of us knew that his shot had told, yet for some reason, perhaps owing to the superior cunning of the birds, none fell at our feet. Before regaining the wagon, however, we found fluttering on our path a fine fat one with a shattered second joint. It was claimed by Sachem, on the ground that in his aiming he had made legs a speciality, not wishing to injure the breasts.

Later in the season, when the birds had become mucn wilder, I often shot them, both running and flying. They are very hard to kill, and a sorely wounded one will often astonish the hunter by running long distances, or hiding where it seems impossible. The fall through the air, or sudden stop from full speed when running, are alike exciting spectacles. And the big body, with red throat and dark plume, luscious even to look at, is fit game to excite the pride of any sportsman.

The modes of hunting the wild turkey are numerous.* Mounted on a swift pony it is not difficult to run one down, as may be done in half an hour, the birds, when pushed, seeking the open prairie and its ravines at once. On foot, with a dog, they can easily be started from cover, and generally rise with a tremendous commotion among the bushes, when they may be brought down with coarse shot. Another method of turkey shooting, and one that became quite a favorite of mine, was to steal out from camp in the gray of early morning—so early that only the tops of the trees were visible against the sky—provided with a rifle and shot gun both. When the birds have once been hunted, extreme caution is necessary to get within seventy yards of them. Upon a high bough, in the gloom, the old gobbler appears twice his real size, looking as long as a rail. Try the rifle first, and, two chances out of three, there is a miss. Then, as the great wings spread suddenly, like dark sails against the sky, and the big body, launched from the bough, shakes the tree top as if a wind was passing through it, catch your shot gun,

* The amateur sportsman or other reader, will find them described at length in the Appendix.

and fire. In the dim light, and at long distance, it takes a quick and trué eye to call from the ground that welcome resound which tells of game fallen.

Under the big oaks, meanwhile, our camp fire burned brightly, and Shamus was developing the mysteries of his art. Roast turkey and broiled antelope tempt the pampered appetites of dyspeptic city men, but here in the wilderness, their fresh juices, hissing from beds of glowing coals, filled the air with a fragrance that to us was sweeter than roses. Tired enough, after an all day's ride, and hungry as bears from twelve hours fasting, we sucked in the odors of the cooking meat, as a sort of aërial soup, while the Dobeen stood an aproned king of grease and turkey, with basting spoon for scepter, and with it kept motioning back the hungry hordes that skirmished along his borders.

Two mess chests had been placed a few feet apart, with the tail-boards of our wagons connecting them, and over this was spread a linen table cloth, white plates, clean napkins, and bright knives, with salt, pepper, and butter. All were in their accustomed places. This our first meal on the plains looked more like an aristocratic pic-nic than a supper in the territory of the buffaloes. But the picture was too bright to last, and ere many days neither napkins nor cloth could have been made available as flags of truce.

It is one of those threadbare truisms, adorning all hunting stories of every age and clime, that hunger is the best seasoning. We had an excess of it on hand just then, and would willingly have shared it with the dyspeptic, baldheaded young men of Fifth Avenue The turkey we found fat and very rich in flavor, and the antelope steaks more delicate than venison. Condensed milk supplied well the place of the usual lacteal, and was an improvement on the city article, inasmuch as we knew exactly what quantity and quality of water went into it. We were obliged to economize, however, respecting this part of our supplies. The following entry in our log-book, by Sachem, under date of the day preceding this, will explain the reason: "Two cans of milk stolen, probably by the Cheyennes. Consider the article more reliable for families than city stump-tail, requiring neither milking or feeding, and never kicking the bucket, or causing infants to do so. Had no idea that a taste of it would develop such a talent for hooking."

CHAPTER XIV.

A CAMP-FIRE SCENE—VAGABONDIZING—THE BLACK PACER OF THE PLAINS—SOME ADVICE FROM BUFFALO BILL ABOUT INDIAN FIGHTING—LO'S ABHORRENCE OF LONG RANGE—HIS DREAD OF CANNON—AN IRISH GOBLIN—SACHEM'S "SONG OF SHAMUS."

HOW vividly, when one is fairly embarked in any new enterprise, do the events of the first night impress one's imagination, and how indelibly do they fix themselves in the memory! Inside our tents all was clean and cheery, but as none of us were disposed to seek them before a late hour, we spent the evening around our camp-fires. Excitement, for the time, had overmastered our sense of fatigue. The Professor's notes were out, and, with his feet to the fire and a box for a desk, he looked more like the Arkansas traveler writing home, than the learned savan committing to paper the latest secrets wrung from nature. The remainder of our party were scattered promiscuously around the fire, some seated on logs and boxes, the others outstretched upon the grass.

Tammany Sachem was the first to break the silence. "Fellow citizens," he exclaimed, "let's vagabondize!" Now, with our alderman, vagabondizing meant story telling, an accomplishment which we consider the especial forte of vagabonds.

We all hailed this proposition gladly, for Buffalo Bill, stretched there before the fire, had much of plain lore stored in his active brain that we wished to draw out, and we at once seized the opportunity to ask about the black pacer we had seen during the afternoon, and his weird story of the bloody saddle.

From Bill's narrative we gathered the following: Something over a year before the era of our expedition a train of government wagons left Fort Hays destined for Fort Harker, and the Indians being troublesome, some twenty soldiers were sent in the wagons, as a guard. A few hours later there passed through Hays City a man from the mountains riding a powerful black stallion, while his family, consisting of a young wife and her brother, occupied a covered wagon which followed close behind. The stranger determined to take advantage of the protection afforded by the government train, and the little party pushed out after it over the plains. The day was a sultry one in midsummer, the sun pouring down its flood of heat on the desolate surface of the expanse that spread away on all sides. The long train, a full mile from front to rear, dragged its slow length sluggishly along, the mules sleepily following the trail, while the teamsters and soldiers dozed in the covered wagons. A driver, who happened to be awake, saw in the distance a beautiful mirage, and in it, as he looked, strange objects, like mounted men, were bobbing up and down. But then he had often seen weeds and other small objects similarly transformed, by these wonderful illusions of the plains, and even he forgot the bobbing shadows and dozed away again on his seat.

But there was danger near. Stealthily out of the mirage, and bending low in their saddles, rode a painted band of savages, hiding their advance in a ravine. Their purpose was to strike and cut off the rear of the train, the length of which promised unusual success to their undertaking, as the white men were too much scattered to oppose any resistance to a sudden onset. At length, nearly the entire train had filed by, and the foremost of the last half dozen wagons approached the ravine. At the signal, out from it burst the troop of red horsemen, and crossed the road like a dash of dust from the hand of a hurricane, every savage spreading his blanket and uttering the war whoop. The startled teams fled in stampede over the plains, dragging the wagons after them. Some of the drivers were thrown out and others jumped. Two or three were killed, and by the time the other teams and the guards had taken the alarm, and turned back for a rescue, the savages had cut the traces of the frightened mules, and were on the return with them to their distant villages. Instead of stopping the animals to release them from the wagons, the Indians urged them to wilder speed, and leaning from their saddles, cut the fastenings at full run. Among the booty taken, was a valuable race horse and fifteen hundred dollars in greenbacks, belonging to an officer who was on his way from New Mexico to the East.

Meanwhile, our friend, the owner of the black pacer, with his outfit, was moving quietly along two or three miles in the rear, entirely unaware of affairs at the front. Some of the savages, while escaping with the booty, espied him, and coveting the noble animal which he rode, they made a detour and surprised him as he sat jogging along a hundred yards or so ahead of the wagon containing his wife and brother-in-law. Though mortally wounded at their first volley, with the desperate effort of a dying man he clung to the saddle for a hundred yards or more, and then rolled upon the prairie a lifeless corpse. Frantic with terror, the horse dashed through the circle of Indians that surrounded him, and fled. The savages, probably fearing longer delay, did not pursue, nor even attack the wagon, and the black pacer was not seen again for some months, when at length some hunters discovered him, freed from saddle and bridle, the leader of the wild herd.

Buffalo Bill gave us quite an insight into some of the mysteries of plain craft. When you are alone, and a party of Indians are discovered, never let them approach you. If in the saddle, and escape or concealment is impossible, dismount, and motion them back with your gun. It shows coolness, and these fellows never like to get within rifle range, when a firm hand is at the trigger. If there is any water near, try and reach it, for then, if worst comes to worst, you can stand a siege. The savages of the plains are always anxious to get at close quarters before developing hostility. Unless very greatly in the majority, and with some unusual incentive to attack, they will not approach a rifle guard. Were they as well supplied with breech-loading guns as with pistols, the case would be different, of course. Bill was the hero of many Indian battles, and had fought savages in all ways and at all hours, on horseback and on foot, at night and in daytime alike.

As an amusing illustration of the savage abhorrence of long-range guns, I beg the reader's indulgence for introducing an anecdote which I afterward heard narrated by an officer who participated in the affair. Major A—— was sent out from Fort Hays with a company of men on an Indian scout, and, when near a tributary of the south fork of the Solomon, the savages appeared in force, and a fight commenced, which continued until dark. Several soldiers were wounded and two killed. As the Indians were evidently increasing in numbers, after nightfall a squad was dispatched to the fort for ambulances and reinforcements. Only six men could be spared, and these were sent off with a light field-piece in charge. Soon after crossing the Saline, a strong band of Indians was discovered half a mile off reconnoitering. A shell was sent screaming toward them, but the aim was too high, and it burst a short distance beyond them. Nevertheless, the effect was instantaneous; the savages vanished, nor stood upon the order of their going. During the next ten miles this scene was repeated three times, the stand-point on each occasion being removed further and further away. The last shot was a remarkably long one, and the shell burst directly in their faces. Not only did they disappear for good, but the whole investing force, on receiving their report, fled likewise.

Talking thus about Indians, under the gloom of the trees, seemed in some unaccountable way to suggest the idea of witches to the mind of Pythagoras. Perhaps, in accordance with his pet theory of development, he was cogitating whether, ages ago, the red man's family horse might not have been a broomstick. At any rate, he suddenly gave a new turn to the conversation by asking Shamus why, when the dogs pointed the witch-hazel during our quail hunt at Topeka, he had affirmed that the canine race could see spirits and witches which to mortal eyes were invisible. Now, the Dobeen had been bred on an Irish moor, where the whole air is woven, like a Gobelin tapestry, full of dreams of the marvelous, and where whenever an unusual object is noticed by moonlight, the frightened peasant, instead of stopping a moment to investigate the cause, rushes shivering to his hut to tell of the fearful *phookas* he has seen. He was very superstitious, and we had often been amused at his evasions, when, as sometimes happened, his faith conflicted with our commands. The time might be near when such peculiarities would prove troublesome instead of amusing, and it was well, therefore, that we should get a peep at the foundations of our cook's faith, and perhaps that portion of it which related to our friends, the dogs, would be especially entertaining. Moreover, we had had so much of the red man that we were glad to welcome an Irish witch to our first camp-fire. Dobeen's narrative was substantially as follows, though I can not attempt to clothe it in his exact language, and still less in the rich brogue which yet clung to him after years of ups and downs in "Ameriky."

"Dogs can study out many things better than men can," said Shamus, in his most impressive manner. "Before I left old Ireland for America, I had a dashing beast, with as much wit as any boy in the country. He could poach a rabbit and steal a bird from under the gamekeeper's nose, an' give the swatest howl of warnin' whenever a bailiff came into them parts."

Sachem suggested that these were rather remarkable habits for a dog connected with the great house of Dobeen.

"But yez must know he was only a pup when my fortunes went by," responded Shamus, "and he learnt these tricks afterward. Ah, but he was a smart chap! Couldn't he smell bailiffs afore ever they came near, an' see all the witches and ghosts, too, by second sight! He wouldn't never go near the O'Shea's house, that had a haunted room, though pretty Mary, the house-girl, often coaxed at him with the nicest bits of meat."

Sachem thought that perhaps the animal's second sight might have shown him that stray shot from pretty Mary's master, aimed at a vagabond, might perhaps hit the vagabond's dog.

"I wasn't a vagabond them times," retorted Shamus, quickly, yet with entire good humor, "and sorry for it I am that the name could ever belong to me since. And please, Mr. Sachem, don't be after interruptin' again. Some people wonder why the dogs bark at the new moon an' howl under the windows afore a death. In the one matter, your honors, they see the witches on a broomstick, ridin' roun' the sky, an' gatherin' ripe moon-beams for their death-mixtures an' brain blights. Many a man in our grandfathers' time—yes, an' now-a-days too—sleepin' under the full moon, has had his brains addled by the unwholesome powder falling from the witches' aprons. Wise men call it comet dust. And why shouldn't a dog that has grown up to mind his duty of watchin' the family, howl when he sees Death sittin' on the window sill, a starin' within, and preparin' to snatch some darlint away? Ah, but their second sight is a wonderful gift though!

"The name of my dog, your honors, was Goblin, an' he came to us in a queer sort of way, just like a goblin should. There was a hard storm along the coast, an' the next mornin' a broken yawl drifted in, half full of water, with a dead man washin' about in it, an' a half-drowned pup squattin' on the back seat. Me an' my cousin buried the man, an' the other beast I brought up. May be there was somethin' in this distress that he got into so young that he could n't outgrow. Even the priest used to notice it, and say the poor creature had a sort of touch of the melancholy; an' sure, he never was a joyful dog. Smart an' true he was, but, faith, he was n't never happy; yez might pat him to pieces, an' get never a wag of the tail for it. He delighted in wakes and buryins, an' when a neighborin' gamekeeper died, he howled for a whole day an' a night, though the man had shot at him twenty times. Mighty few men, your honors, with a dozen slugs in their skin, would have stood on the edge of a man's grave that shot them, an' mourned when the earth rattled on the box the way Goblin, poor beast, did then. Ah, nobody knows what dogs can see with their wonderful second sight. That beast thought an' studied out things better than half the men ye'll find; an' it 's my belief that dogs did so before, an' they have done it since, an' they always will."

"You are right, Dobeen," said the Professor. "Put a wise dog, and a foolish, vicious master together. The brute exhibits more tenderness and thoughtfulness than the man. In the latter, even the mantle of our largest charity is insufficient to cover his multitude of sins, while the skin of his faithful animal wraps nothing but honest virtue. The dog, having once suffered from poison, avoids tempting pieces of meat thenceforward, when proffered by strange hands, but the man steeps his brain in poison again and again—or as often as he can lay hold of it. While grasping the deadly thing, he sees, stretching out from the bar room door, a down grade road, with open graves at the end, and frightened madmen, chased by the blue devils and murder and misery, rushing madly toward them. These swallow their victims, as the hatches of a prison ship do the galley slave, and close upon them to give them up only when the jailer, the angel of the resurrection, shall unlock the tombs, and calls their occupants to judgment. Does the sight appall and bring him to his senses? No, he crowds among the terrors, and takes to his bosom the same venomous serpent that he has seen sting so many thousands to death before him. And yet people give to the brute's wisdom the name of instinct, and call man's madness wisdom."

"But, your honors," interposed Dobeen, "I shall be after losing my dog entirely, unless yez lave off interruptin' me, an' let me finish my story."

"Go on, Shamus, go on!" we all cried with one breath.

"Well, then, when Goblin came to me in his infancy, he wore a silver collar with his name all beautifully engraved on it. May be the dead man in the boat had been bringing him from some strange land to the childer at home, and thinking how the odd name would please them all, when the shadows were darting around his hearth. And so Goblin howled his way through the world, till one full moon eve, when every bog was shinin' as if the peat was silver. Such times, any way in old Ireland, your honors, the air is full of unwholesome spirits. This was good as a wake for Goblin, and I can just hear him now the way he cried and howled that night! He kept both eyes fixed on the moon, and no mortal man, livin' or dead, will ever know what he saw, but when he howled out worse nor common that night, it meant, may be, that some witch, uglier than the rest, had just whisked across the shinin' sky. Just at midnight, I was waked out of a swate sleep by the quietness without, the way a miller is when his mill stops. I looked out of the window at the dog where he sat, an', faith, the dog was n't there at all! Just then I heard a despairin' sort of howl, away up in the air above the trees, an' by that token I knew the witches had Goblin. Next mornin', one of the lads livin' convanient to us told me he had heard the same cry in the middle of the night, the cry, your honors, of the poor beast as the witches carried him off. Afore the week was out, Goblin's collar was found on the gamekeeper's grave; that was all—not a hair else of him was ever seen in old Ireland."

As Shamus concluded his veracious narrative he looked around upon us with an air of triumph, as if satisfied that even Sachem dare not now dispute the second sight of the canine race.

That worthy took occasion to declare on the instant, however, that the nearest neighbor was fully justified in playing the witch. If any thing could destroy the happiness of human beings, as well as of the broom-riding beldams, it would be the howling of worthless curs at night. He himself had often been in at the death of vagabond cats and dogs engaged in moon-worship. The outbursts of Goblin had simply been silenced in an outburst of popular indignation.

SMASHING A CHEYENNE BLACK KETTLE.

CHAPTER XV.

A FIRE SCENE—A GLIMPSE OF THE SOUTH—'COON HUNTING IN MISSISSIPPI—VOICES IN THE SOLITUDE—FRIENDS OR FOES—A STARTLING SERENADE—PANIC IN CAMP—CAYOTES AND THEIR HABITS—WORRYING A BUFFALO BULL—THE SECOND DAY - DAUB, OUR ARTIST—HE MAKES HIS MARK.

OUR fire scene was evidently no novelty to the Mexicans, whose lives had been spent in camping out, and who, with one cheap blanket each, for mattress and covering, slept soundly under the wagons. Across their dark, expressionless faces the flames threw fitful gleams of light, which were as unheeded as the flashes with which the Nineteenth Century endeavors to penetrate the gloom which shrouds them as a nation. While the world moves on, the degenerate descendants of Montezuma sleep.

In the valley bordering our little skirt of trees we could hear the horses cropping the short, juicy buffalo grass, and trailing their lariat ropes around a circle, of which the pin was the center. Semi-Colon lay on the grass close to his father, who occupied a cracker-box seat in this tableau, the amiable son at little intervals raising his head to indorse, in his peculiar dissyllabic way, what the positive parent said. Looking at the group around me, and thinking of our evening turkey hunt, memory carried me back to the last time I had been among the trees after dark, with gun in hand, which was at the South, away down in Mississippi, just after the war.

It was a lazy time, those November days. Large flocks of swans filled the air above, with their flute-like notes, and thousands of sand-hill cranes circled far up toward the sun, their bodies looking like distant bees, as from dizzy heights they croaked their approbation of the rich crops beneath them. Ducks passed like charges of grape shot, sending back shrill whistles from their wings, as they dived down into the standing corn.

As night came on, the moon went up in a great rush of light, like the reflector of a railroad train mounting the sky. Soon every shadow is driven from the woods, and then the horns are tooted, the dogs howl, and away go gangs of woolly heads, old and young, in pursuit of Messrs. 'Possum and 'Coon. In vain the sly tree-fox doubles around stumps, and leaving tempting persimmon and oaks full of plumpest acorns, at the warning noise, seeks refuge among huge cypresses. On go the hunters—big dogs, little dogs, bear-teasers, and deer-hounds, sprinkled with darkeys—crashing through cane and underbrush, the human portion of the party laughing and yelling as if a tempest had stolen them ages ago from Babel, and just discharged them in pursuit of that particular 'coon.

The voice of the Professor suddenly called me back to the present, and I found myself chilled by the wet grass, as if my body had been wandering with the mind in that land of cotton, and was unprepared for the northern air.

"Gentlemen"—this was what the voice said—"we are now one thousand and five hundred miles from Washington City, latitude 39, longitude 99. Stick a pin there on the map, and you will find that we have got well out on the spot that geographers have been pleased to call desert. Does it look like one? Tell me, gentlemen, had you rather discount your manhood among the stumps of New England than loan it at a premium to the rich banks of these streams?"

The Professor came to an abrupt pause, for borne to us on the still air was that most unmistakable of all sounds, the human voice. The note of one bird at a distance may be mistaken for another, and the cry of a brute, when faintly heard, lose its distinguishing tones. But once let man lift up his voice in the solitude, and all nature knows that the lord of animal creation is abroad. There are many sounds which resemble the human voice, just as there are many objects which, indistinctly seen, the hunter's eye may misinterpret as birds. But when a flock of birds does cross his vision, however far away, he never mistakes them for any thing else. The first may have excited suspicion, the latter resolves at once into certainty.

We listened attentively and anxiously. It might very naturally be supposed that, after leaving the abodes of his fellows, and going far out into the solitary places of Nature, man would rejoice to catch the sounds which told him that others of his race were near, but this, like many other things, is modified by circumstances. On the plains the first question asked is, "Are they friends or foes?" No one being able to answer, the breeze and general probabilities are inquired of, and until the eyes pass verdict the moments are laden with suspense. Even in times of peace the hunter, if possible, avoids the savage bands which flit back and forth across Buffalo Land; for, if he saves his life, he is apt to lose an inconvenient amount of provisions, at least, at their hands.

Our guide speedily informed us that Indians never make any noise when in camp, which was gratifying intelligence. All further suspense was shortly relieved by the appearance down the valley of muskets glittering in the moon-light. The bearers proved to be two soldiers, who stated that some officers, with a small force of cavalry, were in camp a mile below us, being out for the purpose of obtaining buffalo meat, and having as guests two or three gentlemen from St. Louis, desirous of seeing the sport. They had heard our late heavy firing, and sent to know what was the matter. We gave the soldiers a late paper to carry back, and with many regrets that our fatigue was too great to think of accompanying them for a neighborly call, we bade them good-night, and saw them disappear down the valley.

At the Professor's suggestion, preparations were now made for retiring, and we sought our tent and blankets. In a few brief moments, the others of the party were blowing, in nasal trumpetings, the praises of Morpheus. I could not sleep, however; for each bone had its own individual ache, and was telling how tired it was. Pulling up a tent-pin, I looked out under the canvas.

On a log by the fire sat Shamus, his head between his hands, gazing at the coals, and droning a low tune. Occasionally, he would make a dash at some fire-brand, with a stick which he used as a poker, and break it into fragments, or toss it nervously to one side. Whether this was because it resolved itself into a fire-sprite winking at him, or some uuhappy memory glowed out of the coals, I tried to tempt sleep by conjecturing.

Off at a little distance, I could see one of our men standing guard near the horses, and once or twice my excited fancy thought it detected shadows creeping toward him. A little beyond, nervously stretching his lariat rope, while walking in a circle around the pin, was Mr. Colon's Iron Billy. His clean head erect, and fine nose taking the breeze, the intelligent animal appeared restless, and I could not help thinking that he saw or smelt something unusual, away in the darkness. What if the bottom grass was full of creeping savages?

The crescent moon, just rising over the divide, was scarred by many cloud lines, and as yet gave no light. The sensation which had stolen over me was becoming disagreeable, when far off, at some ford down the creek, I heard animals splashing through water, and concluded that Billy's nervousness was caused by crossing buffaloes. The horse had an established reputation as a watch, his former owner having assured us that neither Indian nor wild beast could approach camp without Billy giving the alarm.

Presently, Dobeen resumed his droning, which had been suspended for a few moments, this time singing some snatches from an old Irish ballad. The last words were just dying away, when I started to my feet in horror. What an infernal chorus filled the air! Each point of the compass was represented, and we were wrapped around with a discordant, fiendish cordon of sound. Bursting upon us with a deep mocking cry, it ended abruptly in a wild "Ha-ha!" It was such a chorus as pours through Hades, when some poet opens, for an instant, the gate of the damned. Our poor Irishman, at the first sound, had fallen from the log as if shot, but had suddenly sprung to his feet, and was now performing a terror-dance behind the fire with a club. For a moment, I, too, had taken the outburst for the war-whoop of savages, but was saved from a panic by seeing through the gloom the figure of the sentinel still at his post, and the next instant the voice of the guide was lifted, with the re-assuring intelligence—"Only cayotes, gentlemen, only cayotes!"

Mr. Sachem and Mr. Muggs had been lying close behind me in their blankets. The former had given a terrified snort, and then both lay motionless. After the alarm, Sachem admitted that he was frightened. Had always heard that people shot over instead of under the mark in battle. Was resolved to lay low. Had no high views about such things. Muggs had not thought it worth while to get up. Knew they were wolves. Had heard more hextraordinary 'owls before he came to the blarsted country.

But where was the doctor? Echo answered,

"Where?" "Hallo, Doctor!" cried the guide, and a voice from the woods, which was not echo, answered, "Coming!" Again Buffalo Bill lifted his voice in the solitude, and again came an answer, this time in a form of query, "Is it developed, my boy? If so, classify it." And we answered that the birth in the air had developed into wolves, and been classified as the *canis latrans*, noisy and harmless.

Finding that this new lesson in natural history had taken away all desire for sleep, I finished the study by the fire, with our guide for a tutor.

The cayote (pronounced Kī-o-te), in its habits, is a villainous cross between a jackal and a wolf, feasting on any kind of animal food obtainable, even unearthing corpses negligently buried. With the large gray wolf, the cayotes follow the herds of bison, generally skulking along their outskirts, and feeding upon the wounded and outcasts. These latter are the old bulls which, gaunt and stiff from age and spotted all over with scars, are driven out of the herd by the stout and jealous youngsters. Feeding alone, and weak with the burden of years upon his immense shoulders, the old bull is surrounded by the hungry pack. But they dare not attack. One blow of that ponderous head, with the weight of that shaggy hump behind it, is still capable of knocking down a horse. The veteran could fling his adversaries as nearly over the moon as the cow ever jumped, if they only gave him a chance. Like a grim old castle, he stands there more than a match for any direct assault of the army around.

With the tact of our modern generals, a line of in-

MIDNIGHT SERENADE ON THE PLAINS.

vestment is at once formed, and a system of worrying adopted. No rest now for the old bull. He can not lie down, or the beasts of prey will swarm upon him. Again and again he charges the foe, each time clearing a passage readily, but only to have it close again almost instantly. In these resultless sorties the garrison is fast using up its material of war. The ammunition is getting short which fires the old warrior, and sends the black horns, like a battering-ram, right and left among his foes. As long as he keeps his feet he lives, though hemmed in closely by the snapping and snarling multitude. The tenacity of one of these patriarchs is wonderful. For a whole life-time chief of the brutes on his native plains, he has grown up surrounded by wolves. Not fearing them himself, he has easily defended the cows and calves. An attempted siege would once have been but sport to him, and it seems difficult for the brain in the thick skull to understand that Time, like a vampire, has been sucking the juices from his joints and the blood from his veins.

Tired out at length, the old bull begins to totter, and his knees to shake from sheer exhaustion. His shakiness is as fatal as that of a Wall Street bull. As he lies down the wolves are upon him. They are clinging to the shaggy form, like blood-hounds, before it has even sunk to the sod, and the victim never rises again.

The cayotes are very cowardly, and when carcasses are plenty, sleep during the day in their holes, which are generally dug into the sides of some ravine. If found during the hours of light, it is usually skulking in the hollows near their burrows. They have a decidedly disagreeable penchant for serenading travelers' camps at night, so that our late experience, the guide assured me, was by no means uncommon. They will steal in from all directions, and sit quietly down on their haunches in a circle of investment. Not a sound or sign of their coming do they make, and, if on guard, one may imagine that every foot of the country immediately surrounding is visible, and utterly devoid of any animate object. All at once, as if their tails were connected by a telegraphic wire, and they had all been set going by electricity, the whole line gives voice. The initial note is the only one agreed upon. After striking that in concert, each particular cayote goes it on his own account, and the effect is so diabolical that I could readily excuse Shamus for thinking that the dismal pit had opened.

At this point Dobeen approached and cut off my further gleaning of wolf lore. The corners of his mouth seemed still inclined to twitch, showing that the shock had not yet worn off. He was chilled by the night, he said, and did not feel very well, and craved our honors' permission to sleep at our feet in the tent. Consent was given, and as he left us he turned to announce his belief that animals with such voices must have big throats.

It was not yet light, next morning, when our camp was all astir again. Drowsiness has no abiding place with an expedition like ours upon the plains. Should he be found lurking anywhere among the blankets, a bucket of water, from some hand, routs him at once and for the whole trip. Even Sachem, who usually hugged Morpheus so long and late, might that morning have been seen among the earliest of us washing in the waters of the creek.

We were all in excellent spirits, and with appetites for breakfast that would have done no discredit to a pack of hungry wolves. No sign of the sun was yet visible, save a scarcely perceptible grayish tinge diffusing itself slowly through the darkness, and the lifting of a light fog along the creek upon which we were encamped. Although sufficiently novel to most of our party, the scene was quite dreary, and we longed, amid the gloom and chill, for the appearance of the sun, and breakfast. By the way, I have noticed that with excursion parties, whether sporting or scientific, enthusiasm rises and sets with the sun. The gray period between darkness and dawn is an excellent time for holding council. The mind, no less than the body, seems to find it the coolest hour of the twenty-four, and shrinks back from uncertain advances.

Added to the discomforts usually attendant upon camp-life were our stiff joints. The first day upon horseback is twelve hours of pleasant excitement, with a fair share of wonder that so delightful a recreation is not indulged in more generally. The next twenty-four hours are spent in wondering whether those limbs which furnish one the means of locomotion are still connected with the stiffened body, or utterly riven from it; and, if the whole truth must be told, the saddle has also left its scars.

As the edge of the plateau overlooking the river became visible in the growing light, we saw, as on the evening previous, multitudes of buffalo feeding there, and after breakfast a council of war was held. I am somewhat ashamed to record that it voted no hunting that day. To find the noblest of American game some of us had come half away across the continent, and now, in sight of it, the tide of enthusiasm which had swept us forward hitherto stood suddenly still. Not because it was about to ebb, but simply in obedience to certain signals of distress flying from the various barks, and which it was utterly impossible for any of us to conceal.

For mounting a horse was entirely out of the question for that day. Not one of us could have swung himself into saddle for any less motive than a race with death. Our steps were slow and painful, and we felt as if, at this period of life's voyage, every timber of our several crafts had been pounded separately upon some of the hidden rocks of ocean. It was absolutely necessary to go into dock for repairs, and the valley promised to be a pleasant harbor.

It was a truly melancholy spectacle to behold Sachem and Muggs. The liveliest and the gayest ones yesterday, but to-day the gravest of the grave. That rotund form, which always doubted his own or other people's emotions, was the walking embodiment of woe, and for once evidently clear of all doubt upon one subject, at least. Muggs was even free to confess that, for general results, yesterday's rough riding exceeded "a 'unt with the 'ounds." Our animals were also quite stiff, but the hostlers attributed this not so much to their yesterday's service as to their long ride in the cars. They had not yet got their "land legs" fully on again. It was soothing to our pride, if not to our feelings, to reflect that perhaps some of our soreness was the result of their first day's stiffness.

A beaver colony near us, and a great abundance of turkeys, offered lessons in natural history of no small interest, and within reach of lame students. The valley gave an entomological invitation to Mr. Colon, and the great ledges, with their possibilities of valuable fossils, attracted the Professor.

Sitting on a wagon tongue, and applying liniment to an abraded shin, might have been seen Pythagoras, M. D., whose daily life, since leaving Topeka, had been a series of struggles with the brute he rode. His belief in the transition of souls into horses was growing upon him. He felt that he was combating the spirit of a deceased prize-fighter, which used its hoofs as fists, landing blows right and left. Doctor David called these "spiritual manifestations." A favorite habit of the animal was what is known as brushing flies from the ear with the hind foot, and often, as the owner was about to mount, this species of front kick would upset him. The equine's disposition, it must be said, had not been improved by the immense saddle-bags with which the Doctor had surmounted him when on the march. Originally, these contained a small amount of medicine, but this had all been ground to powder under the weight of sundry stones and bones, gathered in the furtherance of the great theory of development.

As the sun got well up in the heavens, staying in camp became monotonous, aud we hobbled off in different directions, to examine the surroundings. Our

Mexicans climbed to the plains above, taking their rusty muskets along to kill buffalo. Our guide went down to the hunting camp below us, intending to return to Hays with the officers, home duties requiring his attention. One of our hostlers, familiar with the country, was to be our pilot in future.

Back of our camp lay the castellated rocks which had attracted our notice the previous evening, and over which Daub, our artist, now became intensely enthusiastic. He wandered back and forth in front of them, his soul in his eyes, and these upturned to the bluffs. And thus we left him.

"Genius is struggling hard for utterance there," said the Professor impressively. "That young man will make his mark; see if he does n't." Alas, how little we thought he would do it so soon.

An hour later, returning that way, we descried our artist high up on the face of the rocks, perched on a jutting fragment, and clinging to a stunted cedar with one hand, while with the other he plied his brush. Fully forty feet intervened between him and the earth.

"What devotion!" cried the Professor.

"Beautiful spirit," said Mr. Colon, "how soon it commences to climb."

"That young man will develop," said Dr. Pythagoras.

A few feet more, and the artist and his work were fully revealed. He had developed. A cry of agony came from the Professor's lips; for there in large yellow lines, half blotting out a beautiful stone, our eyes beheld the diabolical letters, S O Z.

He never finished the word. The Professor seized a rifle, and brought it to a level with the artist's paint pot. "Come down, you rascal!" he cried. "How dare you deface one of nature's castles with a patent name?" Would he have fired? I think he would. But the man of genius caught his eye, and comprehending the situation, cried, with face whiter than the chalk before him, "O, don't!"

"Add the 'odont', you villain," screamed the Professor, "and I'll—I'll fire!"

With our first returning wagon, the artist went back to Hays, but his work, alas! remains, and perhaps—who knows?—some future generation may yet point to that wall and tell how SOZ, king of an extinct people, once held dominion over the beautiful valley.

CHAPTER XVI.

BISON MEAT—A STRANGE ARRIVAL—THE SYDNEY FAMILY—THE HOME IN THE VALLEY—THE SOLOMON MASSACRE—THE MURDER OF THE FATHER AND THE CHILD—THE SETTLERS' FLIGHT—INCIDENTS—OUR QUEEN OF THE PLAINS—THE PROFESSOR INTERESTED—IRISH MARY—DOBEEN HAPPY—THE HEROINE OF ROMANCE—SACHEM'S BATH BY MOONLIGHT—THE BEAVER COLONY.

AT noon we were all in camp again, fully prepared to do justice to the ample dinner of buffalo, antelope, and turkey which we found awaiting us. The Mexicans brought in the quarter of an old bull, and, according to their own story, had committed terrible slaughter on the plain above; but, as we had already learned to balance a Mexican account by a deduction of nine-tenths for over-drafts, we felt that we saw before us the result of their day's hunt. This our first taste of bison, gave us highly exaggerated ideas of that animal's endurance. The entire flesh was surprisingly elastic—indeed, a very clever imitation of India rubber. It recoiled from our teeth with a spring, and just then I should scarcely have been surprised had I seen those buffalo which were feeding in the distance, go bounding off like immense foot-balls. My opinion in regard to buffalo meat afterward underwent a great change, but not until I had tasted the flesh of the cows and calves. Shamus, on this occasion, had devoted his culinary energies especially to the turkeys, and they were well worthy such attention. Their fat forms, nicely browned, would have tempted the veriest dyspeptic.

Just as we rose from dinner, a covered emigrant wagon was discovered approaching us, coming down the valley right on our trail. From the fact that we were off the route of overland travel, our first conjecture was that it was from Hays, with a party of hunters, or possibly with Tenacious Gripe, so far recovered as to be rejoining us. We assumed an attitude of dignified interest, prepared to develop it into friendship, or "don't want to know you" style, as occasion might require. A hale, elderly man was the driver, now walking beside his oxen. The outfit halted before our astonished camp, and as it did so two women, genuine spirits of calico and long hair, lifted a corner of the wagon cover and looked out. Both were apparently young, but one face was thin, and had that peculiar expression of being old before its time which is far more desolate than age. The other countenance was certainly good-looking and interesting—quite different, indeed, from those usually seen peeping out of emigrant wagons. Introductions are short and decisive on the plains. We liked their looks, and invited them to stop; they liked ours, and accepted. I think the Professor's dignified attitude and scholarly bearing stood us in good stead as references.

Another female developed as the wagon gave forth its load—this time a bouncing Irish girl, rosy-cheeked and active, evidently the family servant. At this latter apparition Shamus dropped one of our platters, but quickly recovering himself, began to put forth wonderful exertions to prepare a second dinner, the new comers having consented, after some hesitation, to become our guests during the nooning hour.

Before proceeding to give the reader the history of this interesting family, I ought, perhaps, to say that I do so with their express permission, the only disguise being that, at his request, the father will here be designated by his Christian name, Sydney.

These people, after an absence of about a year, were now returning from Elizabeth City, a recently-started mining town in New Mexico, to their former home, about forty miles east of our present camp, which they had left the preceding season under circumstances that were sad, indeed. About three years before, the family, then consisting of Mr. Sydney and wife, and their two daughters, had moved from Ohio to Kansas and settled on a tributary of the Solomon. Availing himself of the homestead law, Mr. Sydney took a tract of one hundred and sixty acres, and commenced improving it. One of the daughters soon married a young man to whom she had been betrothed at the East, and who at once set earnestly to work to make for himself and young wife a home in the new land. The houses of the father and the child were but half a mile apart, and, no timber intervening, each could be plainly seen from the other. For a time this little colony of two families was very happy. Having had the first choice, their farms were well situated, embracing both river and valley, and their herds, provided with rich and unlimited range, increased rapidly. Soon rumors came from below that a railroad, on its way to the Rocky Mountains, would shortly wind its way up the Solomon Valley, bringing civilization to that whole region, and daily mails within a few miles of their doors.

The second year of prosperity had nearly ended, when one morning a man from the settlements above dashed rapidly past Mr. Sydney's house, turning in his saddle to cry that the Cheyennes had been murdering people up the river, and were now sweeping on close behind him. The message of horror was scarcely ended when the dusky cloud appeared in sight, rioting in its tempest of death down the valley. Midway between home and the house of her daughter, Mrs. Sydney was overtaken by the yelling demons. In vain the agonized husband pressed forward to the rescue, firing rapidly with his carbine. She was killed before his eyes, but not scalped, the Indians evidently considering delay dangerous.

It is a fact that speaks volumes in illustration of the mingled ferocity and cowardice that characterize the wild Indians of to-day that, in all that terrible Solomon massacre, not a single armed man who used his weapon was harmed, nor was one house attacked. The victims were composed entirely of the surprised and the defenseless, overtaken at their work and on the roads.

Passing the dead body of the mother, the Cheyennes, on their wiry ponies, swept onward, like demon centaurs, toward the home of the daughter. Sitting by our fire at evening, with that dreary, fixed look which one never forgets who has once seen it, the young woman told us the story of her childless widowhood. Her face was one of those which, smitten by sorrow, are stricken until death. Once evidently comely, the smiles and warm flush had died out from it forever—just as in the lapse of centuries the colors fade from a painting. Though scarcely twenty-five, her youth was but an image of the past. She told her story in that mechanical, absent sort of manner which showed that no morning had followed the evening of that desolate day. She was still living with her dead.

"The Lord gave me then a cup so bitter," she said, "that its sting drove a mother's joy from my heart forever. I have been at peace since, because, among the dregs, I found that God had placed a diamond for me to wear when I was wedded to him. Even then I did not rebel and reproach my Maker, but I sunk down with one loud cry, and it went right along to the great white throne up there, with the spirits of my husband and my babe. I thought I could see them in the air, like two white doves flitting upward, bearing with them, as part of our sacrifice, the cry that I gave, when my heart-strings seemed to snap, and I knew that I was a widow and childless. Perhaps I was crazed for a moment, or—I do not know—perhaps my spirit really did go with them part of the way. The neighbors found me there for dead, and I remained cold, till they brought in my dear babe, my poor, mutilated babe, and placed him on my breast. His warm blood must have woke me, and I sat up, and saw them bringing John's body to lay it by me. And then the whole scene came before me again, and it seemed so stamped into my very brain, that shutting my eyes left me more alone with my murdered ones and the murderers. And I just dragged myself where I could look at the setting sun, and tried with its bright glare to burn the scene from off my vision, so that, if I went mad, there would n't be any memory of it left. For mad people have their memories and suffer from them, and they know it, and the very fact that they know it keeps them mad. I went through it all.

"A person dreaming is not rational, and yet may suffer so, and feel it too, as to shudder hours after waking up. There was John, running toward the house with our baby boy, and the savages yelling and whipping their ponies, trying to get between the open door and him. Alone, he could have saved himself. And our baby thought John was running for play, and was clapping his little hands and chirping at me as the savages closed around my husband. I had only time to pray five words, "O God, save my husband!" and it did not seem an instant until I saw the poor body I loved so well lying on the ground, and they standing over, shooting their arrows into it. Baby was not killed, but thrown forward under one of the horses, and I had just taken a step or so toward him, when an Indian, who seemed to be the chief, lifted him by the dress to his saddle. I think his first intention was to carry him with them, but, seeing some of our neighbors hurrying toward us, they struck the baby with a hatchet, and hurled him to the ground. At the instant they struck him, he was looking back at me with his great blue eyes wide open and staring with fright."

And then the poor woman, having finished hei story, began sobbing piteously.

The Solomon had numberless tales of these terrible massacres equally as harrowing as this, and I could fill pages of this volume with chapters of woe that terminated many a family's history. The result of these and other Indian atrocities is probably yet remembered throughout the entire country. Kansas well nigh rebelled against a government which left her unprotected. The War Department authorized vigorous measures, and the Governor of the State raised a regiment and at its head took the field. Through blows from Custar and Carr, the savages found out, at last, that the dogs of war which they let loose might return to bay at their own doors.

Two women from the Saline were carried into captivity by the Indians, and taken as wives by two of their chiefs. One day Carr, at the head of his troops, looked down into the valley upon the encampment of a band especially noted for its hostility, now lying in fancied security below him. The two white captives were in the wigwams. Suddenly, to the ears of the savages, came a murmur from the hill-side like the first whisper of a torrent.

Instantly, almost, it increased to a roar, and, as they sprung to their feet and rushed forth, the blue waves of vengeance dashed against the village, and broke in showers of leaden spray upon them. Mercy put no shield between them and that annihilating tempest. Every savage in the number was a fiend, and, as a band, they had long been the scourge of the border. Their hands were yet red with the blood of the massacres upon the Saline and Solomon, and white women toiled in the wigwams of their husbands' murderers. One of the captives, Mrs. Daley, was killed by the savages, to prevent rescue; the other was saved, and restored to her husband.

Somewhat later, two women from the Solomon were taken captive, one of them being a bride of but four months who had recently come out with her young husband from the State of New York. Custar seized some chiefs and, with noosed lariats dangling before their eyes, bade them send and have those prisoners brought in, or suffer the penalties. Indians have an unconquerable prejudice against being hung, as it prevents their spirits entering the happy hunting grounds, and the captives were promptly sent to Custar's camp. We afterward saw one of them, Mrs. Morgan, on the Solomon. What an agony must have been hers, as she came in sight of her old home, and the memory of her wrongs since leaving it, rose anew before her!

But to return to the history of our emigrants. After the murders, Mr. Sydney and his daughters abandoned their farms, and with the same wagon and oxen which two years before had brought the family out from Ohio, they started for the recently discovered mines in New Mexico. The journey was tedious, and, when at length arrived there, he found but little gold, and even less relief from his mighty sorrow. The old home, with its graves, beckoned him back, and thither he was now returning to spend his remaining days, unless, as he laconically stated, some one had "jumped the claim." Lest my readers toward the rising sun should not clearly understand the old gentleman's meaning, I ought perhaps to explain that, under existing laws, a "Homesteader" can not be absent from his land over six months at any time, without forfeiting his title, and rendering it liable to occupancy by other parties. It was already two days over the allotted period, he said. But the oxen were thin, and he finally decided to rest with us until the next morning, and then push forward.

Flora, the younger daughter, was a blooming Western girl of a thoroughly practical turn, and a counselor on whose advice the father and sister evidently relied greatly. The Professor assured me confidentially that evening, and with much more than his wonted enthusiasm on such a subject, that she preferred the language of the rocks to that of fashion plates. She had even disputed one of his statements, he said, and vanquished him by producing the proof from a well-worn scientific work—one of a dozen books carefully wrapped up and stowed away with other goods in the wagon.

A novel accomplishment which the young lady possessed was that of being an excellent rifle shot, and it afforded us all considerable merriment when she challenged Muggs to a trial of skill, and, producing a target rifle, utterly defeated him. Such a woman as that, the Professor said, was safe on the frontier; she could fight her own way and clear her vicinity of savages, whenever necessary, as well as any of us.

We did not wish our emigrant maiden aught but

what she was, and were well pleased with the romance of her visit. For the nonce, she was our queen; the rough ox-wagon was her throne, and the great plains her ample domain. In sober truth, she might justly challenge our esteem and admiration. Here was one of the gentler sex willing to make divorce of happiness, that she might minister to a half-crazed father and mourning sister, and who, for their sake, chose to wander through a country which might at any moment become to them the valley of the shadow of death. In the presence of such heroism, what right had we, though bruised and tired, to complain? No wonder the Professor took early occasion to tell us that she was a noble woman, an honor to her sex.

This emigrant wagon, with its wee bit of domestic life, was a pleasant object to all of us out there on the desert, with the single exception of Alderman Sachem. That worthy member of our party avoided its vicinity, as if a plague spot had there seized upon the valley. "I did think," he exclaimed, dividing glances that were quite the reverse of complimentary between the Professor and Shamus—"I did think that we had got out of the latitude of spooning. We have n't had a digestible mouthful since they came in sight. A love-struck Irishman can neither eat, himself, or let others."

But Shamus was too happy to heed the remark, for the first time since starting, he seemed perfectly contented. An Irish girl, the like of Mary, and devoted enough to follow her old master through such adversity, seemed Dobeen's beau ideal of the lovely and lovable in the sex. The valley became for him the brightest spot upon earth. He would have been content there to court and cook, I think, during the remainder of his natural life. Mary was shy, and Shamus was bold, but it was quite apparent that both enjoyed the situation immensely.

Although the little party stayed but a day, their departure seemed to leave quite a void in the valley. The most noticeable results to us were some errors in cooking and a slackness in the prosecution of scientific investigations.

Mr. Sydney gave us a hearty invitation to visit him upon the Solomon, if our wanderings took us that way, and our prophetic souls, with a common instinct, told all of us that the Professor would recognize a call of science in that direction. By a look and a smile from a maiden, the Philosopher, deeply sunken in the primary formation, had been drawn to the surface of the modern, a result which fashionable society had more than once striven in vain to bring about. Miss Flora certainly bid fair to become a favorite pupil of his, were the opportunity only offered.

This maiden of the plains was a new character. The beautiful heroine mentioned in most Western novels as having penetrated the Indian country, is either the daughter of "once wealthy parents," or the heiress of a noble family and stolen by gypsies for reward or revenge. It was the first appearance that I could recall of a farmer's girl in a position where kidnapping Indians and a frantic lover could so easily appear, and by opportune conjunction weave the plot of a soul-harrowing romance.

Another evening in camp was spent in writing and

story-telling. The fire was getting low, when Sachem rose to his feet and called to Shamus. "Dobeen," said he, "your country folks are always handy with the sticks. Let 's go for wood, and have a fire that will warm up the witches on their broomsticks and send them flying off to hug the clouds." We watched the pair go out of sight. Knowing well the habits of Tammany, we all felt sure that, though he might find the load, Irish shoulders would have to bear it back to camp.

Scarcely three minutes had elapsed, when out of the timber, with garments as wet as water could make them and dripping fast, a fat form came shivering to our fire. Our alderman had taken a night bath in the creek—an adventure which he thus related in his own peculiar way:

"Below us in the woods is a big beaver pond, I don't know how deep. I seemed an hour going down, and did n't touch bottom then. I was fooled by the moon. (To be expected, though, as she 's a female!) A few of her beams, thrown down through the trees, glittered on the water like drift wood. That sort of beams make poor timber for bridges, but I didn't know it then as well as I do now. One of them went from bank to bank, and I took it for a log, and got a ducking. How frightened I was, though, when my feet touched water and my body went, with a swash, right under it! I opened my mouth to shout and the water rushed in, and I was like a vessel sinking with open hatches. I took in so much, I was afraid I'd be waterlogged and never come up. I did, though, and found that rascally Irishman throwing sticks at my head, and telling me to hold on to them. I told him to do that thing himself, and finally climbed ashore."

We afterward sought out our newly-found neighbors, the beavers, finding their pond a short distance below us on the creek, and a little lower down the dam itself. Many more trees had been cut for the latter than were used in its construction, several having been abandoned when almost ready to fall. We noticed that the butts of the prostrated trees were sharpened down gradually like the point of a lead-pencil, but both ways, instead of one, so that a tree cut nearly through met from above and below at the point of breaking, like the waist of an hour glass. This dam was most interesting to all of us, since it seemed so much to resemble the work of man. In this waste place of the earth, it really seemed almost like company, and we felt a strong desire to have a friendly conference with the builders. But these had formed this resorvoir for the express purpose that in its depths they might escape intrusion, and now the whole regiment of engineers seemed asleep in barracks. Still our men secured a few very fine ones by trapping.

It appeared that the beavers were a vacillating set of architects, as all the trees which stood near the water and leaned over it at all, were gnawed more or less, and many of them left when almost ready to fall. The position of the dam had evidently been determined by the tree which fell first. From the reckless manner in which they had slashed around with their teeth, it was pertinently suggested that this colony must have obtained from the beaver congress a government subsidy. Having been acquainted with the art of building before man mastered it, the beaver race also probably understood how to do it at little personal expense.

The beaver appears to be distributed in considerable numbers all over the western half of Kansas, although the spring floods sweep away their dams almost every season. Once afterward, when lost on the plains for a day, I came across a beaver dam. Several hours of anxious suspense in the solitude, fearing to meet man lest he should prove a savage, begot a strange feeling of companionship when I came in sight of the rude structure of logs. If not civilization, it was a close imitation of it, and I laid down and fell into a refreshing sleep, soothed, in the fantasies of Dreamland, with the whir of looms and hum of factory life.

CHAPTER XVII.

PREPARATIONS FOR THE CHASE—THE VALLEY OF THE SALINE—QUEER 'COONS—A BISON'S GAME OF BLUFF—IN PURSUIT—ALONGSIDE THE GAME—FIRING FROM THE SADDLE—A CHARGE AND A PANIC—FALSE HISTORY AGAIN—GOING FOR AMMUNITION—THE PROFESSOR'S LETTER—DISROBING THE VICTIM.

THE early dawn of Wednesday morning saw us again astir. There was the same creeping of mist out of the valley to join the darkness as it fled from the plains above, and the same revealing of thousands of shaggy forms silently feeding in the distance. This time our beasts and our bodies were both in excellent condition for the chase. Joints gain and lose stiffness quickly in such a life. One morning the hunter feels as if the mill of life, though he turn its crank ever so slowly, had broken every bone in his body; twenty-four hours later may find him elastic and buoyant, as if youth had torn away from the embrace of the dead past and was with him again in all its pristine vigor. In the present case, too, that friend of early hours and foe of sleepy eyes, the coffee bean had done its work for us grandly.

Ten horsemen comprised the strength of the party which rode out of the valley just as daylight was coming into it. One of the hostlers and a Mexican were left in camp, the remainder of our force accompanying us, with a couple of wagons to bring in the game. At his earnest solicitation, Shamus was permitted to abandon his post of duty temporarily, and go along also, with the understanding that he was to select choice pieces from the first suitable game we might bring down, and, returning to camp, be ready for our arrival with an ample dinner.

As we rode down the valley of Silver Creek, gangs of wild turkeys occasionally came out of the narrow skirt of timber, and, running along before us for short distances, re-entered it, and were lost to view again. Never having been hunted, they seemed destitute of the timidity and cunning which are the usual characteristics of this bird.

Twenty minutes' ride brought us to the Saline, the basin of which we found to be half a mile or thereabouts in width, and presenting a scene of great desolation. We were something like two hundred feet below the table-lands which came down to the narrow valley in barren canyons and masses of rock. The stream itself is narrow, with less than two feet of water running swiftly over the sands, and along its banks, at intervals, a few dwarfed cottonwood trees. Such was the Valley of the Saline at this point; yet thirty miles below, our men told us, the valley opened out into rich bottom lands, and was famous for its beauty.

While in the act of crossing, we came suddenly upon four small animals playing and fishing in the shallow water. With an exclamation of astonishment, the Professor had his glasses out in a moment. The guide informed us they were only 'coons, and such they were sure enough, with the peculiar color and distinctive rings that made it impossible, on second look, to mistake them for any thing else. Truly, Nature seemed full of eccentricities in this remarkable region. The raccoons of natural history have always affected trees, and been considered, *par excellence*, creatures of the forest. I scarcely think the Professor would have been surprised, at that moment, to know that hereabouts fish were in the habit of climbing around in bushes, or stealing corn.

When they heard us, the four little fellows scampered away a few steps, and disappeared in some holes in the bank, in executing which maneuver one of them swam a yard or two across a deep spot, making good progress. We learned from our men that small colonies of these animals are frequently found along treeless creeks on the plains, living in the banks, and fishing for a living, by grasping the minnows and frogs, as they pass over the shallow places.

From the river we directed our course toward a deep canyon which, opening toward us as if the bluff had been riven asunder by some great convulsion of Nature, at its further end reached the level of the plains, and offered us an easy ascent. Evidence of volcanic action appeared along the canyon in the form of vitrified fragments and occasional masses of lava resembling rock.

The guide called our attention to an object in the ravine some distance ahead, which was enveloped in a cloud of dust. It was a buffalo, he said, indulging in a game of bluff. This statement not appearing very clear to our non-gambling party, he explained that the old fellow was "butting against the bank, as if he was going to break it all to pieces, when in reality he had no show at all."

As we could not approach nearer without frightening him, we stood still for a few minutes and watched him. He would back fifteen or twenty yards from the bluff, paw the ground for an instant, and then fling himself headlong against the wall of earth with a tremendous force, as was abundantly testified by the great clouds of dust that would rise in the air. For a moment afterward he would continue violently hooking the soil, as if the bowels of the earth were those of an adversary. We afterward repeatedly saw bulls engaged in this exercise. It is to the buffalo what the training school is to the prize-fighter, a developing of brute force for future conflicts.

The shock of such charges as we witnessed, if made by a domestic ox, would have broken his neck. Even our bison friend finally overdid the matter. Either because his foot tripped or the blow glanced, upon one of his charges, he fell down on his fore legs, and then rolled completely over. We thought this a good time to push forward, and accordingly did so at a gallop. Whether thinking himself knocked down by a foe, or because he heard the rattling of hoofs, we could not determine, but he suddenly sprang to his feet, whirled his shaggy head into bearing upon us, then turned and set away at full speed up the canyon, toward the plains above. The order was given to ply spur and close in upon him, if possible, or he would set the herds above in motion.

It was a mad ride that we had for the next ten

minutes—across beds of gravel, among huge bowlders, and once or twice over great fissures in the earth which chilled my blood as I took a sort of bird's-eye view of their depths. In a lumbering run on ahead of us went the frightened bull, his feet occasionally sending back dashes of pebbles, while behind him rattled such a clattering of hoofs that the poor brute, if he could think at all, must have imagined he had butted open the door of Hades, and was now being pursued by its inmates.

There were mishaps in this our first buffalo hunt, of course, and among them, Muggs dropped a stirrup, and was obliged to support himself afterward on one foot—an awkward matter, resulting from his inconvenient English saddle, one of the kind which compels one, half the time, to sustain the whole body by the stirrups alone. We gained upon the game steadily, though no particular member of our party excelled as leader, first one being ahead and then the other. Cynocephalus developed wonderfully, and kept well up with his better conditioned neighbors.

What a magnificent prize for the hunter rushed on before us, swinging his ponderous head from side to side, for the purpose of getting better rear views—such an ungainly and shaggy animal, a perfect marvel of magnificent disproportions! It is well enough to go to Africa and hunt lions, and describe their majestic, flowing manes; but this bison, in mad flight ahead of us, could have furnished hair and mane enough to fit out half a dozen lions. At close quarters, too, he was fully as dangerous as the king of beasts.

We were close at his heels when the level of the plain was reached, and pursuer and pursued shot out upon it together. A large herd, feeding not five hundred yards away, was speedily in full flight northward. "A stern chase is a long chase," is no less true in buffalo hunting than in nautical matters. After considerable experience in the sport, I would recommend amateurs to get as near their game as possible before starting, and then try their horses' full metal. Once by the side of the game, he can keep there to the end. And so, after a terrible chase, when at times we had almost despaired of overtaking the old fellow, we now found it easy to keep alongside.

Our bull was a huge one, even among his species, and in such moments of excitement the imagination seems to have a trick of entering the chambers of the eye, and sliding its mirrors into a sort of double focus arrangement. With blood boiling until my heart seemed to bob up and down on its surface, I found myself riding parallel with the brute, and had I never seen him afterward, would have been almost willing to make oath that his size could be represented only by throwing a covering of buffalo robes over an elephant.

Every one in the party was firing, some having dropped their reins to use their carbines, and others yet guiding their horses with one hand, while they fired their holster revolvers with the other. Shooting from the saddle, with a horse going at full speed, needs practice to enable one to hit any thing smaller than a mammoth. You point the weapon, but at the instant your finger presses the trigger, the muzzle may be directed toward the zenith or the earth. An experienced hunter steadies his arm, not allowing it to take part in the motion of his body, no matter how rough the latter may be. But we were not experienced hunters, and so, although such exclamations as, "That told!" "Mine went through!" and "Perfectly riddled!" were almost as numerous as the bullets, it was easy to see that the flying monster remained unharmed.

From the first, Mr. Colon had fired without taking any aim whatever, and so it happened that his gun, in describing its half circle consequent upon the rising and falling motion of the horse, at length went off at the proper moment, and we heard the thud of the ball as it struck. Dropping his head into position as if for a charge, the buffalo whirled sharply to the right, and passing directly between our horses, made off toward the main herd. But he soon slowed down to a walk, and as we again came up with him, we could see the blood trickling from his nose, which he held low like a sick ox.

In the excitement of the chase, and perhaps from being well blown before coming near the buffalo, our horses had hitherto shown no fear, but now, as the old bull stood there in all his savage hugeness, and the smell of blood tainted the air, they pushed, jostled, snorted, and pranced, so that it required all our efforts to keep them from downright flight. Even Dobeen's donkey kept his rider uncertain whether his destiny was to seek the ground or abide in the saddle.

The brute stood facing us, perhaps fifty yards off, his eyes rolling wildly from pain and fury, and the blood flowing freely through his nostrils.

We were waiting patiently for him to die, when suddenly the head went into position, like a Roman battering ram, and down he came upon us. We were utterly routed. No spur was necessary to prompt the horses, and I doubt if their former owners had ever known what latent speed their hides concealed. The whole thing was so sudden there was no time for thought, and all that I can remember is a confused sort of idea that each animal was going off at a tremendous pace, with the rider devoting his energies to sticking on. After the first few jumps, we were no longer an organized company, each brute taking his own course, and carrying us, like fragments of an explosion, in different directions. A marked exception, however, was Muggs' mule, which for the only time in his life, seemed unwilling to run away. After being the first to start, and assisting the others to stampede, he stopped suddenly short, depositing his rider something like ten yards ahead of him, in a manner quite the reverse of gentle.

We did not stop running as soon as we might have done. And I here enter protest against the nonsense indulged in on one point by most of the novelists who educate people in buffalo lore. When we halted, there stood the bull not thirty yards from the spot where he had first stopped, although we had located him, throughout more than half a mile's ride but a few feet from our horses' tails, and at times had even imagined we heard his deep panting. This mortifying record would have been saved us had we known that a buffalo's charges never extend beyond a short distance. Either his adversary or his attack is speedily terminated. He does not pursue, in the "long, deep gallop" style at all. Yet I scarcely remember a single instance mentioned in those old books of western adventure, in which a buffalo's charge was for a less distance than a mile. In one case that I now recall, the race was nip and tuck between man and bison for over an hour, and the biped was finally en abled to save his life only by leaving the saddle and swinging into a tree! Such stories are simply balderdash.

As soon as possible after checking our horses, we rode back toward the wagon and the game, seeing in the former, the grinning faces of our men. The buffalo was still on his feet, but while we looked he slowly sunk to his knees, like an ox lying down to rest, and then quietly reposed on his belly, in the same attitude one sees domestic cattle assume when wishing a quiet chew of the cud. Had it not been for his bloody nose and wild eyes, he would have looked as peaceful as any bovine that ever breathed.

Wishing to put the poor brute out of misery, we approached closer, and several of us dismounted, when a general fire was opened. Like a cat, the old fellow was on his feet again almost instantly. By a singular coincidence, our entire party just then discovered that we were out of ammunition, and in a body started for the wagon, to get some. Muggs afterward assured us that, at the time, he had just got his hand in, "so that every shot told, you know," and

GOING AFTER AMMUNITION.

I have the authority of all for the deliberate statement that the bull would have been riddled before moving a foot had not the cartridges suddenly given out.

The effort of getting up had sent the mass of blood collected from inward bleeding surging out of the buffalo's nose, and, as we looked back, he was tottering feebly, and an instant afterward fell to the ground. There was no doubt now of his death, and we swarmed upon and around him. He was an immense old fellow, and his hide fairly covered with the scars of past battles. Inasmuch as this was our first trophy, it was determined to take his skin, and we forthwith seated the Professor on his great shaggy neck, with the horns forming arms for an impromptu hunter's throne. From thence he wrote upon leaves from his note-book a letter to his class at the East, which he permitted me to copy. I introduce it here, as showing that the blood of even a savan pulsates warmly amid such circumstances as now surrounded us.

"ON A BUFFALO, IN THE
YEAR OF MY HAPPINESS, ONE.

"*Dear Class*—I know the staid and quiet habits that characterize all of you, and that you are not given to hard riding and buffalo hunting. Yet this prairie air, with its rich fragrance and wild freeness, would give a new circulation to the blood of each one of you. Like a gale at sea, the breeze sweeps against one's cheeks, and the great billows of land rise on every side, as mountains of troubled ocean. Why not desert the city and lose yourself for awhile in this great grand waste? Antelope are bounding and buffalo running on every side of us, while villages of prairie dogs bark at the flying herds. One grows in self-estimation after breathing this air, and, feeling that safety and life depend on his own exertions, learns to place reliance upon the powers which Nature has given him, with manly independence of artificial laws and police.

"While I am writing, the first victim of our prowess, a magnificent specimen of the American bison, is being skinned by our suite, the robe from which, when prepared, we intend sending you. The men say it must be dressed by some of the civilized Indians on the reserves, as the white man's tanning injures the value.

. . . .

"The robe is now off, and half a ton of fat meat lies exposed. We shall only take the hind quarters, a portion of the hump, and the tongue. How glad the famishing wretches in the tenement houses of the city would be for an opportunity to pick those long ribs which we leave for the wolves! His horns are somewhat battered, but we have cut them off, to supplant hooks on a future hat-rack. One of the men has just taken a large musket ball from the animal's flank. That shot must have been received years ago, as the ball is an old fashioned one and is thickly encased in fat.

"The geological formation of the country is very interesting. I expect to examine the same more thoroughly after we have studied the animals traversing its surface. Yesterday, we had in camp a family from the Solomon, who were sufferers some months since from the fearful Indian massacre there. Their story was an exceedingly interesting one, though very sad. We shall visit them if duty calls that way. I must close. The men have thrown the skin in the wagon, flesh side up, and deposited the meat upon it, and all are now ready for further conquests.

"Your sincere friend and instructor,

"H——."

CHAPTER XVIII.

STILL HUNTING—DARK OBJECTS AGAINST THE HORIZON—THE RED MAN AGAIN—RETREAT TO CAMP—PREPARATIONS FOR DEFENSE—SHAKING HANDS WITH DEATH—MR. COLON'S BUGS—THE EMBASSADORS—A NEW ALARM—MORE INDIANS—TERRIFIC BATTLE BETWEEN PAWNEES AND CHEYENNES—THEIR MODE OF FIGHTING—GOOD HORSEMANSHIP—A SCIENTIFIC PARTY AS SEXTONS—DITTO AS SURGEONS—CAMPS OF THE COMBATANTS—STEALING AWAY—AN APPARITION.

OUR further conquests for that day, it was decided, could best be effected by still hunting. The guide had suggested that, if we desired to fill our wagon with meat and get back to camp before night, we might profitably adopt the practice of old hunters, who, when they pursue bison, "mean business." The new tactics consisted of infantry evolutions, and required a dismounting of the cavalry. We were to crawl up to the herds, through ravines, and from those ambuscades open fire.

A mile away buffalo were feeding in large numbers, and our men pointed out several swales into which we could sink from the surface of the plains, and, following the winding lines, find cover until emerging among the herd. But while we were still gazing at the latter, sharp and distinct against the northern horizon appeared other objects, evidently mounted men, and men in that direction meant Indians. It is wonderful how quickly one's ardor disappears, when, from being the hunter, he becomes the hunted. Our only desire now was, in Sachem's language, "a hankering arter camp," which we at once proceeded to gratify.

Back again with the remainder of our party, we felt quite safe. Indians of the plains seldom attack an armed body which is prepared for them; and then there had been no recent demonstrations of hostility. On the other hand, no massacre had yet occurred upon the frontier which was not unexpected. The whole life of many of these nomads has been a catalogue of surprises. It was Artemus Ward, I think, who knew mules that would be good for weeks, for the sake of getting a better opportunity of kicking a man. These savages will do the same for the sake of killing one.

Many an armed man, fully capable of defending himself, has thus been thrown off his guard, and sent suddenly into eternity. The cunning savage, seeing his foe prepared, approaches with signs of friendship, and cries of "How, how?"—Indian and short for "How are you?" Their extended hands meet, and as the palms touch, the pale-face shakes hands with death; for, while his fingers are held fast in that treacherous clasp, some other savage brains him from behind, or sheaths a knife in his heart, and the betrayed white, jerked forward with a fiendish laugh, kisses the grass with bloody lips. We had been repeatedly warned by our guides that, when in the minority, the only safe way to hold councils with the Indians is at rifle range. Even if bound by treaty, a knowledge that they can take your scalp without losing their own, is like binding a thief with threads of gold: the very power which should restrain, is in itself a temptation.

Our little camp soon bristled all over with defiance, a sort of mammoth porcupine presenting points at every angle for the enemy's consideration. Our animals were put safely under cover among the trees, where they could not be easily stampeded; the wagons were ranged in a crescent, forming excellent defense for our exposed side; and pockets were hurriedly filled with ammunition. As we were thus earnestly preparing for war, an entomological accident occurred. Sachem, while excitedly thrusting a handful of cartridges into Mr. Colon's pockets, suddenly drew back his hand with an expression of alarm, bringing with it a whole assortment of bugs. One of the pocket-cases of our entomologist had opened, and the inmates, imprisoned but that morning, were now swarming over our fat friend's fingers, and up his arm, which he was shaking vigorously. There they were—rare bugs and plethoric spiders, together with one lively young lizard—all clinging to the limb which had brought them rescue from their cavernous cell with more tenacity than if they had been stuck on with Spalding's glue. Poor Sachem! While he danced and fumed, and gave his opinion of bug-men generally, Mr. Colon cried—"O, my bugs, my beautiful bugs!" and grasped eagerly at his vanishing treasures. Our alderman disengaged himself at length from his noxious visitors, and meanwhile the other members of the party, having provided themselves, poured into the other pocket of the grieved naturalist a further supply of cartridges, thereby utterly annihilating the remainder of his collection.

Our preparations being concluded, and still no signs of the Indians, we sat down to dinner. Shamus was terribly agitated, and the shades of dyspepsia hovered over his cooking; but, although the coffee was muddy and the meat burned, we were in no mood to take exceptions. There was considerable determination visible on the faces of all our party. The red man was getting to be as sore a trouble to us as the black man had been to politicians, and having already lost a day on his account, we were now fully resolved to hold our ground. We had seen the savage in all the terrors of his war-paint, and felt a very comforting degree of assurance that a dozen cool-headed hunters, mostly armed with breech-loaders, possessed the odds.

At length, along the edge of the breaks beyond the Saline, a dark object appeared, followed by another and then another in rapid succession, until forty unmistakable Indians came in sight, and were bearing directly toward us, following the tracks of our wagons. Half a mile off they halted, and then we saw one big fellow ride forward alone. His form seemed a familiar one, and soon it revealed itself as that of our late friend, White Wolf. Now we had, but a few days before, in the space of four brief hours, concluded at least forty treaties of peace with this chief and his drunken braves; yet, remembering past history, we should have wanted at least as many more treaties, before taking the chances of having one of them kept, and admitting the painted heathens before us to full confidence and fellowship.

As the leader of our party, it devolved upon the Professor to go forward and meet the chief, which he promptly did, taking along our man who was acting in Cody's place as guide, to assist him in comprehending the savage's wishes. Midway between us the respective embassadors met. We heard the chief's loud "How, how?" and saw their hand-shaking, and could not help wondering what the Philosopher's class would say, could they have beheld their honored tutor officiating as a frontispiece for such a savage background.

White Wolf stated that he had been out after Pawnees; he could not find them, and so "Indian felt heap bad!" Just at this instant a loud, quick cry came from his knot of warriors, who were now manifesting the wildest excitement, lashing up their ponies, stringing their bows, and making other preparations as if for a fight. Without a word, the chief turned and ran for dear life toward his band, while the Professor and our guide wheeled and ran for dear life toward us. Seldom has the man of science made such progress as did the respected leader of our expedition then. The guide called, "Cover us with your guns!"—a command which we immediately proceeded to obey, evidently to the intense alarm of the Professor, for so completely were they covered, that I doubt if either would have escaped, had we been called upon to fire.

Our first thought had been a suspicion of treachery,

but we now saw that the Cheyennes had faced toward the hills, and, following their gaze, we beheld coming down their trail, and upon the tracks of our wagon, another band of mounted Indians. It soon became clear to us that the Pawnees, the Wolf's failure to find whom had made that noble red man feel "heap bad," were coming to find him. We counted them riding along, twenty-five in all—inferior in numbers, it was true, but superior to the Cheyennes in respect to their arms, so that, upon the whole, the two forces now about to come together were not unevenly matched. The Pawnees live beyond the Platte, and for years have been friendly to the whites, even serving in the wars against the other tribes on several occasions.

What a stir there was in the late peaceful valley! The buffalo that were lately feeding along the brow of the plateau had all fled, and here right before us were sixty-five native Americans, bent upon killing each other off, directly under the eyes of their traditional destroyer, the white man. The Professor said it forcibly suggested to his mind some of the fearful gladiatorial tragedies of antiquity. Sachem responded that he was n't much of a Roman himself, but he could say that in this show he was very glad we occupied the box-seat, the safest place anywhere around there; and we all decided that it must be a face-to-face fight, in which neither party dare run, as that would be disorganization and destruction.

It was strange to see these wild Ishmaelites of the plains warring against each other. Over the wide territory, broad enough for thousands of such pitiful tribes, they had sought out each other for a bloody duel, like two gangs of pirates in combat on mid-ocean; and, like them, if either or both were killed, the world would be all the better for it. It was clearly what would be called, on Wall street, a "brokers' war," in which, when the operators are preying on each other, outsiders are safe.

While we were looking, a wild, disagreeable shout came up from the twenty-five Pawnees, as they charged down into the valley, which was promptly responded to by fierce yells from the forty Cheyennes.

"Let it be our task to bury the dead," said the Professor, looking toward the wagon in which rested his geological spade. "It is extremely problematical whether any of these red men will go out of the valley alive."

And thus another wonderful change had come over the spirit of our dream. From being a scientific and sporting expedition, we had been suddenly metamorphosed into a gang of sextons, who, in a valley among the buffaloes, were witnessing an Indian battle, and waiting to bury the slain.

As the Pawnees came down at full gallop, the Cheyennes lashed up their ponies to meet them. Then came the crack of pistols, and a perfect storm of arrows passed and crossed each other in mid-air. As the combatants met, we could see them poking lances at each other's ribs for an instant, and then each side retreated to its starting point. Charge first was ended. We gazed over the battle-field to count the dead, but to our surprise none appeared.

A few minutes were spent by both parties in a

BUREAU OF ILLUSTRATION BUFFALO

BATTLE BETWEEN CHEYENNES AND PAWNEES.

general overhauling of their equipments, and then another charge was made. They rode across each other's fronts and around in circles, firing their arrows and yelling like demons, and occasionally, when two combatants accidentally got close together, prodding away with lances. The oddest part of the whole terrible tragedy to us was that the charges looked, when closely approaching each other, as if they were being made by two riderless bands of wild ponies.

The Indians would lie along that side of their horses which was turned away from the enemy, and fire their pistols and shoot their arrows from under the animals' necks, thus leaving exposed in the saddle only that portion of the savage anatomy which was capable of receiving the largest number of arrows with results the least possibly dangerous. I noticed one fat old fellow whose pony carried him out of battle with two arrows sticking in the portion thus unprotected, like pins in a cushion. He still kept up his yelling, but it struck me that there was a touch of anguish in the tone, and I felt confident that he would not sit down and tell his children of the battle for some time to come.

We saw one exhibition of horsemanship which especially excited our admiration. An arrow struck a Cheyenne on the forehead, glancing off, but stunning him so with its iron point, that, after swaying in the saddle for an instant, he fell to the earth. Another of the tribe, who was following at full speed, leaned toward the ground, and checking his pony but slightly, seized the prostrate warrior by the waistband, and, flinging him across his horse in front of the saddle, rode on out of the battle.

For several hours—indeed until the sun was low in the heavens and the shadows crept into the valley—this terrible fray continued, the charging, shouting, and firing being kept up until both combatants had worked down the river so far that we could no longer see them.

It was approaching the dusk of evening when White Wolf and his band rode back. We counted them and found the original forty still alive. The chief assured us they had killed "heap Pawnees," whereupon some of us sallied forth to visit the battle-field. Three dead ponies lay there, and with a disagreeable sensation we looked around, expecting to discover the mangled riders near by. Not one was visible, however, nor even the least sign of their blood. The grass was not sodden with gore, nor did a single rigid arm or aboriginal toe stick up in the gathering gloom. Neither the wolves or buzzards gathered over the field, and slowly the conviction dawned upon us that Indian battles, like some other things, are not always what they seem.

As we turned again toward camp, the Professor, dragging his spade after him, suggested that, in accordance with the reputed habits of these savages, the Pawnees had perhaps carried off their dead. But at the instant, only a short distance down the river, the camp-fire of that miserable and all but annihilated band glimmered forth. It was decidedly too bold and cheerful for the use of twenty-five ghosts, and we knew then that White Wolf had lied.

That valorous chieftain we found limping around outside our wagons, with a lance-cut in one of his legs, while several of his warriors had arrow-wounds, and one a pistol-shot, none of the injuries, however, being dangerous. The Pawnees probably suffered with equal severity; and this was the sum total of the day's frightful carnage—the entire result of all the fierce display that we had witnessed.

Not long afterward, in front of a Government fort, and in plain sight of the garrison, a battle occurred between two large parties of rival tribes, about equal in numbers. Back and forth, amid furious cries and clouds of arrows, the hostile savages charged. Noon saw the affair commenced, and sunset scarcely beheld its ending. The Government report states, if my memory serves me correctly, that one Indian and two horses were killed; and a shade of doubt still exists among the witnesses whether that one unlucky warrior did not break his neck by the fall of his pony!

These savages fight on horseback, and are neither bold nor successful, except when the attacking party is overwhelming in numbers, and then the affair becomes a massacre. All this knowledge came to us afterward, but our first introduction to it was a surprise. Kind-hearted man though he was, I think the resultless ending of the battle disconcerted even the Professor. Having nerved one's self to expect horrors, it is natural to seek, on the gloomy mirror of fate, some rays of glimmering light which can be turned to advantage. I think the Professor's rays, had the contest proved as sanguinary as we first antici-

15

pated, would have found their focus in some stout cask containing a nicely-pickled Pawnee or Cheyenne *en route* to a distant dissecting table. It would have been rather a novel way, I have always thought, of sending the untutored savage to college.

We made a requisition upon our medicine-chest, and dressed the wounds of the suffering warriors. White Wolf stripped to the waist, and, exposing his broad, muscular form, exhibited thirty-six scars, where, in different battles, lances and arrows had struck him. It struck us all as a rather remarkable circumstance, though we prudently refrained from commenting upon it just then, that nearly all these scars were on his back.

The chief expressed great friendship for us, and I really believe he felt it. Sachem's stout form was especially the object of his admiration. Between these two worthies a very cordial regard seemed to be springing up, until White Wolf unluckily offered him an Indian bride and a hundred buffalo robes, if he would go with the band to its wigwams on the Arkansas—a proposition which disgusted our alderman beyond measure. Savages, sooner or later, generally scalp white sons-in-law, and it would be "heap good" for the Cheyenne to have such an opportunity always handy. Sachem declined the honor with all the dignity he could command, and carefully avoided "the match-making old heathen," as he termed him, for the remainder of the evening.

We kept early hours that night. Guard was doubled, to prevent any possible treachery, and a sleepy party laid down to rest. The Cheyennes went into camp a few hundred yards up the creek, a barely perceptible light, looking from our tents like a fire-fly, marking the spot.

When a "cold camp" is discovered on the plains, the experienced frontiersman can always determine at once whether white men or Indians made it, by the size of the ash-heap. The former, even when trying to make their fire a small one, will consume in one evening as much fuel as would last the red man a half-moon. The latter, putting together two or three buffalo chips, or as many twigs, will huddle over them when ignited, and extract warmth and heat enough for cooking from a flame that could scarcely be seen twenty yards.

The two opposing parties, which were now resting only a mile or so apart, had each tested the other's metal, and, as the sequel proved, found them foemen worthy of their *steal.* From the unconcealed fires in their respective camps, we concluded that neither side had any intention of attacking, or fear of being attacked.

It was early in the dawn of the next morning when we were startled from our slumbers by a terrific cry from Shamus, which brought all of us to our tent-doors, with rifles in hand ready to do battle, in the shortest possible time. Looking out, we beheld our cook standing near the first preparations of breakfast, and gazing with astonished eyes toward the darkness under the trees, among which we heard, or at least imagined we heard, the stealthy steps of moccasined feet. In answer to our interrogatories, Shamus stated that just as he was putting the meat in the pan, he saw the light of the fire reflected, for an instant, on a painted face peering out at him from behind a tree. "Faith, but I shaved the lad's head wid the skillet!" said Dobeen, and sure enough we found that article of culinary equipment lying at the foot of the suspected cottonwood, badly bent from contact with something, but whether that something was the bark or a painted skull is known only to that skulking Cheyenne.

We waited until broad daylight, but no further disturbance occurred, and what was strangest of all, the valley both above and below us seemed entirely destitute of either Pawnee or Cheyenne. A reconnoissance, which was made by the Professor, Mr. Colon, and our guide, developed the fact that not being able to steal any thing else, the savages had executed the difficult military maneuver of stealing away. Just before daybreak, the Pawnees had gone due north, and the Cheyennes, about the same time, due south. As White Wolf had expressed a cold-blooded intention of exterminating the remnant of his foes in the morning, the pitying stars may have taken the matter in hand and misled him; and if so, how disappointed that blood-thirsty band must have been when their path brought them into their own village, instead of the Pawnee camp! In confirmation of this astrological suggestion, I may say that while in Topeka I saw "stars," on several occasions, leading Indians in the opposite direction from that in which they wished to go.

In due time our party sat down to another plentiful breakfast, which was eaten with all the more relish because we had all that little world to ourselves again. Discussing Dobeen's apparition, we finally came to the unanimous conclusion that it was some Indian who, while his brothers stole away, had straggled behind, to pick up a keepsake. I think that hideous face among the trees never entirely ceased to haunt the chamber of Dobeen's memory. He shied as badly as did Muggs' mule, when in strange timber, and was ever afterward a warm advocate for pitching camp on the open prairie.

In justice to White Wolf, it should be stated that we afterward learned that while charging in such a mistaken direction after Pawnees that morning, he met two men from Hays City, out after buffalo meat. Finding that they were from the village which had been kind to him, he loaded their wagons with fat quarters, instead of filling their bodies with arrows, as they had first expected, and sent them home rejoicing.

CHAPTER XIX.

STALKING THE BISON—BUFFALO AS OXEN—EXPENSIVE POWER—A BUFFALO AT A LUNATIC ASYLUM—THE GATEWAY TO THE HERDS—INFERNAL GRAPE-SHOT—NATURE'S BOMB-SHELLS—CRAWLING BEDOUINS—"THAR THEY HUMP"—THE SLAUGHTER BEGUN—AN INEFFECTUAL CHARGE—"KETCHING THE CRITTER"—RETURN TO CAMP—CALVES' HEAD ON THE STOMACH—AN UNPLEASANT EPISODE—WOLF BAITING, AND HOW IT IS DONE.

BREAKFAST over, the day's work was planned out. We were desirous of loading one of our wagons with game, and sending it back to Hays, from whence the meat could be forwarded by express to distant friends, and serve as tidings from camp, of "all's well." The other wagon we decided to keep with us. Horseback hunting, although fine sport, evidently would not, in our hands, prove sufficiently expeditious in procuring meat. Our guide adduced another argument as follows: "Yer see, gents, if yer want ter ship meat by rail, it won't do ter run it eight or ten miles, like a fox, and git it all heated up. Ther jints must be cool, or they 'll spile." Stalking the bison was to be our day's sport, therefore, and we were speedily off, taking only the two wagons, the riding animals being all left in camp. Shamus prepared a lunch for us, as we did not expect to return for dinner before dusk.

Following the same route as the day before, we

soon ascended the Saline "breaks," and emerged on the plains above. Looking to us as if they had not changed position for twenty-four hours, the buffalo herds still covered the face of the country, busy as ever in their constant occupation of feeding. For animals which perform no labor, they have an egregious appetite, eating as if they were Nature's lawn-gardeners, and were under contract with her to keep the grass shaved.

What an immense aggregate of animal power was running to waste before us. Those huge shoulders, to which the whole body seemed simply a base, were just the things for neck-yokes. Others, indeed, had thought the same before us, and tried to utilize these wild oxen. A gentleman at Salina, Kansas, obtained two buffalo calves, and trained them carefully to the yoke. They pulled admirably, but their very strength proved a temptation to them. A pasture-fence was no obstacle in the way of their sweet will. Not that they went over it, but they simply walked through it, boards being crushed as readily as a willow thicket. In summer they took the shortest road to water, regardless of intervening obstructions, and they thought nothing of flinging themselves over a perpendicular bank, wagon and all. After carefully calculating the result of his experiment at the end of the first year, the owner decided that, although he undoubtedly had a large amount of power on hand, he could obtain a similar quantity, at less expense, by buying a couple of steam-engines.

A few months previous to our trip, a contractor on the Kansas Pacific Railroad determined to domesticate a young bison bull, and accordingly took it to his home at Cincinnati. Proving a cross customer, he presented it to the Longview Lunatic Asylum, near that city, but there was no inmate insane enough to occupy the yard simultaneously with Taurus for any length of time. The first day he charged among the lunatics in a reckless manner, eliciting surprising activity of crazy legs. If exercise for their minds was what the poor creatures needed, they certainly obtained it, by calculating when and where to dodge.

Without loss of time, we set about finding a gateway into the herds. Looking at the surface before us, it appeared a level, unbroken plain, quite to the verge where it rolled up against the distant horizon. One would have maintained that even a ditch, if there, might be traced in its meanderings across the smooth brown floor. Yet deep ravines, miles in length, wound in and out among the herds, though to us entirely invisible. A short search discovered one of these, which promised to answer our purpose, and to lead to a spot where a large number of cows and calves were feeding. Fortunately the wind was north, so that we could creep into its teeth without sending to the timid mothers any tell-tale taint.

The wagons were stopped, and we got out, and descending into the hollow, moved forward. The walls on either side seemed disagreeably close. All around us was animal life, a small portion of which would have been sufficient, if so disposed, to make the concealed path which we were traversing a veritable "last ditch" to us. As we entered the ravine, some cayotes slunk out of it ahead of us, and one large gray wolf, with long gallop, disappeared over the banks. The temptation to fire at them was very strong, but prudence and the guide forbade.

We picked up some very fine specimens of "infernal grape," in the form of nearly round balls of iron pyrites. They lay upon the surface like canister-shot upon a battle-field. It seemed as if during the early period, when Mother Earth began to cool off a little, her fiery heart still palpitated so violently under her thin bodice, that beads of the molten life within, like drops of perspiration, had forced their way through, and, in cooling, had retained their bubble-like form. We could have picked up a half-bushel of them which would have made very fair aliment for cannon. The dogs of war could have spit them out as spitefully and fatally against human hearts as if the morsels had been prepared by human hands. From such well-molded shot, of no mortal make, Milton might have obtained his charges for those first cannon which the traitor-angel invented and employed against the embattled hosts of heaven. Shamus, when he afterward became acquainted with the specimens, called them "a rattlin' shower of witches' pebbles."

We also passed large surfaces of white rock, which were sprinkled all over with dark, hollow balls, of a vitrified substance. Most of them were imbedded midway in the rock, leaving a hemisphere exposed which, in color and form, was an exact counterpart of a large bomb. If the reader has ever seen a shell partly imbedded in the substance against which it was fired, this description will be perfectly plain.

There were indications that a volcano had once existed in this vicinity, and it seemed highly probable that the red-hot balls which it projected into air had fallen and cooled in the soft formation adjacent, still retaining their original shape.

We should have lingered longer over these geological curiosities, had not the premonitory symptoms of a scientific lecture from the Professor alarmed our guide into the remonstrance, "You're burnin' daylight, gents!" and thus warned, we pushed forward.

A few hundred yards further brought us to the spot for commencing active operations. Dropping upon hands and knees, we began crawling along the side of the ravine in a line, pushing our guns before us. We knew that the buffalo must be very close, for we could hear the measured cropping of their teeth upon the grass. They seemed to be feeding toward us, as we slowly drew up to the level. I found myself trembling all over, so nervous that the cracking of a weed under our guns sounded to me as loud as a pistol-shot.

I looked around, and the stories which I had read in my youth of adventures in oriental lands rose fresh to my memory. I almost imagined our party a dozen wild Bedouins, creeping from ambush to fire upon a caravan, the first note of alarm to which would be a storm of musketry. Unshaven faces, soiled clothes, and rough hair, assisted us to the personation, and if aught else was needed to carry out the fancy, it soon came in a low "Hist!" from the guide, as he pointed to the level above us. Following the direction of his finger, we saw some hairy lumps, about the size of muffs, not fifty yards in front of us, bobbing up and down just above the line which defined the prairie's edge against the sky. For an instant, we supposed them to be small animals of some sort, playing on the slope, but the low voice of the guide said, "Thar they hump, gents!" and we caught the word at once, just as the whaler does the welcome cry of "There she blows," from the look-out aloft. What we saw, of course, were the humps of buffaloes moving slowly forward as they fed. At a word from our guide, we halted for last preparations.

"Fire at the nearest cows, gents," he said, "and if you get one down, and keep hid, you'll have lots of shots at the bulls gatherin' round."

Muggs was continually getting his gun crosswise, so that should it go off ahead of time, as usual, it would shoot somebody on the left, and kick some one on the right. Just ahead of us, a prairie dog sat on his castle wall, and barked constantly. But, fortunately, neither his signals nor our grumbled remonstrances to the Briton seemed to attract the attention of the herd in the least degree.

A few more feet of cautious crawling, and several buffaloes stood revealed, a cow and calf among the number. The mother espied us, and lifting her uncouth head, with its crooked, homely horns, regarded us for an instant with a quiet sort of feminine curiosity, and then went to feeding again. She probably considered us a parcel of sneaking wolves, and being conscious of having hosts of protectors near her, was not at all frightened. Almost simultaneously, the guns of the whole party were at shoulder, and just as the cow lifted her head again, to watch the movement, we fired. The fate of that bison was as effectually sealed as that of the condemned army horse which was first used to tell Paris and the world the terrors of the mitrailleuse. The poor creature gave a quick whirl to the right, made two convulsive jumps, and then stood still. She dropped her nose, a gush of blood following fast; her whole frame shuddered, as the air from the lungs tried to force its way through the clotted tide, and then she fell dead, almost crushing the calf also. The smell of the blood seemed to excite the bulls more than the report of the guns, which had only startled them for an instant. Some stood stupidly snuffing about the prostrate victim, while others, straightening out their tails, marched uneasily around.

Lying on the ground, and our heads only visible, we kept up a constant firing. It was almost impossible not to hit some of the old bulls. The veterans were wounded rapidly, and in all portions of their bodies. One old fellow, who had been standing with his rear to us, suddenly took it into his head to run for dear life, and away he went accordingly, with his hams looking very much like the end of a huge pepper-box. Two or three others soon began to show signs of grogginess, being drunk with the blood which was collecting internally from their many wounds.

One bulky and distressed specimen suddenly

caught a glimpse of the Professor's hat. Forthwith the tail was straightened and raised stiffly into the air, the head was lowered, and down he came upon us at full charge. Such a proceeding, a few days before, would simply have resolved itself into a question whether he could catch us or not. Now, however, we stood our ground, or rather we lay upon it very firmly, while enough of us took careful aim to batter his bones fast and sorely. Before taking twenty steps, he was limping from a shattered foreleg, and in a moment more came to a sullen halt, and shook his old head in impotent rage. His eyes were fixed fiercely upon ours; he evidently desired nothing in the world so much as to get forward for a closer acquaintance, but his broken bones forbade. We fired rapidly, and fairly loaded his body with lead before he allowed death to trip him from his feet. He never took his eyes from off us, until the body rolled over, and I thanked our breech-loaders which had prevented the poor beast from having a fair chance.

Three buffalo were down, as the result of our first "stalk." The herd had fled, but the calf we had first seen remained standing stupidly by his dead mother. "Let 's ketch the critter," said our guide, and to catch him we accordingly prepared. The first movement was to surround him, which done, we began closing in upon him. He was hardly larger than a good-sized goat, and we feared might succeed in dodging us, but as the circle narrowed, our hopes of securing a live specimen increased. Suddenly, the little fellow seemed aware of his danger, and, whirling about, with head down, made a dart for the open space between Sachem and the guide. As they closed to prevent his escape, our fat friend went down with a butt in the stomach, which, although far from pleasant, was nevertheless the occasion of sufficient delay on the part of the calf to enable the guide and Semi-Colon to lay firm hold upon him. It was wonderful what a warlike little fellow he proved, butting undauntedly at our legs, and uttering, as he did so, a hissing noise. "But me no butts," exclaimed the Professor, with a facetiousness which from him was almost as amusing to the rest of us as the pugnacity of the calf, as he sprang aside to avoid a blow on the knee, and suddenly recognized Duty's call in another direction. It was not long, however, before the little animal was securely bound, and laid in one of the wagons, which by this time had come up.

The work of skinning and cutting up our game now began, the robe of the cow proving finer than that from either of the others. Our men told us that from one position old hunters sometimes shoot down a dozen buffalo before the herd takes flight. Success is much more probable if the first victim is a female.

Other herds invited our attention, and by three o'clock in the afternoon we had twenty quarters secured, and were returning to camp. Only the first three robes had been taken off, the skin being left on the rest of the meat, the better to preserve it from soiling.

Such hunting fatigues one, and we were glad enough to see the smoke of our fire rising from the valley, and to anticipate the dinner which we felt was waiting for us. The plains tired us, and so did conversation, and all instinctively felt that any attempt at a joke, in our hungry, worn out condition, would have caused an all but fiendish state of feeling. Momus himself could not have made that party smile. Most of us had taken part in cutting up the carcasses, and as we now rode home, sitting on the skin-covered quarters, we looked like a party of butchers returning from the slaughter-pens.

As we drew close to camp, how goodly a sight did Shamus seem, in his white apron, bidding us "Hurry to yer dinner!" while backing up his invitation were the brown turkeys, the stews and roasts, the white bread and yellow butter, and a clean table-cloth. On the spot, we could have pardoned Shamus all his notions of witchcraft, and I think that Sachem's charity just then would even have covered our cook's late weakness in the line of "spooning." The Professor's science, Colon's philanthropy, Sachem's wealth of worldly wisdom, and Muggs' British self-complacency, all combined, offered no such consolation, in this hour of sober realities, as the simple Irishman, with his basting-spoon.

Water from the brook and towels from the chest soon removed blood and dust, and dinner followed. Shamus had many a mark scored against Sachem for attacks on himself and his ancestry, and ventured during dinner to rub out one, by asking Tammany, in a very respectful manner, and as if it was a matter of our *cuisine*, whether calves' heads agreed with his stomach.

What would have been called in Washington, "an unpleasant episode," was discovered by Muggs in the center of a biscuit. Taking a hearty British bite from it, various hairy lines followed the morsel into his mouth, and caught among his teeth. Examination revealed one of Mr. Colon's choicest spiders, which by some means had effected his escape and crawled into the dough. It was hard to tell which was most incensed, the Briton or the entomologist. Sachem remarked that the specimen was much kneaded, and added it to our bill of fare as "game, breaded."

As night approached, our Mexicans prepared for wolf-baiting. During the day they had shot two or three old bulls, which wandered within half a mile of camp, and now the swarthy fellows intended to turn an honest penny. For these purposes professional hunters, and occasionally teamsters on the plains, provide themselves with bottles of strychnine, and a quantity of this was accordingly produced. We went with the men to see the operation, as it clearly came within the province of our studies. With their knives the Mexicans cut from the carcass lumps of flesh about the size of one's fist, into which gashes were made, doses of strychnine inserted, and the flesh then pressed together again. The balls, thus charged, were scattered close around the carcass, and a few laid upon it. Cuts were also made, and the poison introduced in various parts of the hams. As many as fifty doses were thus prepared, and we then returned to camp.

No cayote serenade occurred that night, the musicians evidently being busy drawing sweetness from the cords of the slain. A solemn hush lay over the land, for the bisons are a quiet race, and, except in novels, never take to roaring any more than they do to ten-mile charges.

CHAPTER XX.

THE CAYOTES' STRYCHNINE FEAST—CAPTURING A TIMBER WOLF—A FEW CORDS OF VICTIMS—WHAT THE LAW CONSIDERS "INDIAN TAN"—"FINISHING" THE NEW YORK MARKET—A NEW YORK FARMER'S OPINION OF OUR GRAY WOLF—WESTWARD AGAIN—EPISODES IN OUR JOURNEY—THE WILD HUNTRESS OF THE PLAINS—WAS OUR GUIDE A MURDERER?—THE READER JOINS US IN A BUFFALO CHASE—THE DYING AGONIES.

THE next day's life began, as did the previous one, before sunrise, and while breakfast was cooking, we followed the Mexicans down to examine their baits. The ground around the carcasses was flecked with forms which, in the early light, looked like sleeping sheep. A half-dozen or more wolves, which were still feeding, scampered away at our approach. From the number of animals lying around, we at first supposed most of them simply gorged, but the rapid, satisfied jabbering of the Mexicans quickly convinced us that the strychnine had been doing its work more effectually than we had given it credit for. Twenty-three dead wolves were found, and the even two dozen was made up by a large specimen of the gray variety—or timber-wolf, as it is called in contradistinction from the cayote—who was exceedingly sick, and went rolling about in vain efforts to get out of the way.

Before proceeding to skin the dead wolves, the

Mexicans captured this old fellow and haltered him, by carbine straps, to the horns of one of the buffalo carcasses, near which he sat on his haunches, with eyes yellow from rage and fright. Just to stir him up, we tossed him a piece of bone; he caught it between his long fangs with a click that made our nerves twitch. Man never appreciates the wonderful command that God gave him over the other animals until away from his fellows, and surrounded by the wild beasts of the solitudes, in all their native fierceness. Here were a few mortals of us encompassed by wolves, in sufficient numbers and power to annihilate our party, and yet one solitary man walking toward them would have put the whole brute multitude to flight.

Although we wondered, at the time, that so many wolves were gathered from a single baiting, we soon learned that this success was by no means unusual. At Grinnel Station, where a corporal's guard was stationed, we afterward saw over forty dead wolves, and most of them of the gray variety, stacked up, like cord-wood, as the result of one night's poisoning by the soldiers.

The remainder of this day was devoted to stalking, and resulted in our obtaining a sufficiency of robes and meat to justify us in sending the two Mexican wagons back with them to Hays. Our two captives, the buffalo calf and wolf, went also. The history of that shipment merits brief chronicling.

The robes went to St. Louis, to a man who advertised a patent way of curing such skins, "warranted as good as Indian tan." Some months afterward they were returned to Topeka, duly finished, and I find in the official note-book the following entry. "Robes received to-day. Resolution, by the company, to learn what the law would consider 'Indian tan,' in a suit for damages." They had been shaved so thin that the roots of the hair stuck out on the inside, while the patent liquid in which they had been soaked gave forth an odor which would have been wonderful for its permanency, if it had not been still more wonderful for its offensiveness.

Of the meat, a portion went to our friends, and the balance to Fulton Market, New York. In the first quarter, it carried dyspepsia and disgust, and was so tough that the recipients, with the utmost effort, could not find a tender regret for our danger in obtaining it; while our New York consignee wrote that the first morning's steaks "finished the market," and very nearly finished his customers. He found it impossible, even by the Fulton Market method of subtraction, to get three hundred dollars' worth of express charges out of half that amount of sales, and suggested a discontinuance of shipments. The buffalo calf died on the cars, which probably saved somebody's bones from being broken in celebration of his maturity. The gray wolf got safely to the State of New York, but escaping soon after, a county hunt became necessary, to save the sheep from total extinction. One farmer, in his ire, even went so far as to threaten us with a suit for violating the law, and importing a pauper and disreputable character into the State.

Our experience may be useful to future hunters, to all of whom we would say, unless solely to find amusement, never kill old bulls. Cows and calves are generally juicy and tender, but not so the veterans; they, after death, butt around among one's digestive organs with a ferocity which makes the liver ache. Being most easily obtained, bull beef is generally all that is sent to market, and thus many a patriarchal bison, dead, accomplishes more in retaliation for his sudden taking-off than the Fates ever permitted him to do in lusty life.

A few days more were spent in our Silver Creek camp, and we then folded our tents and took a westward course, with the purpose of examining, not only the remoter regions of Kansas, but also the Colorado portion of the plains. The new town of Sheridan, fourteen miles east of the State line, and nine from Fort Wallace, was our objective point.

"Gentlemen," said the Professor, as we packed and adjusted our things in the wagons, "we are now to climb for a hundred miles directly up the roof of the Rocky Mountain water-shed, its long rivers and rich valleys forming the gutters, or spouts, to carry off the surplus water."

Sachem, who dreaded these lectures almost as much as he did crinoline, interposed with some of his usual badinage; but among irreverent classes of Sophomores and Freshmen, the Professor had learnt to answer only such questions as were relevant, and to pass all others by unheeded. For this reason such interruptions never broke the thread of his discourse, and but temporarily checked its unwinding. In a few minutes, however, the wagons started, and our expedition began crawling up the slope of the Professor's metaphorical roof, and thereupon our worthy leader's discourse was brought to a graceful conclusion.

For four days we continued our westward journey, the soft grass carpet beneath us ever stretching away to the horizon in its tiresome sameness, its figures of buffalo and antelope, big antlered elk and skulking wolves woven more beautifully upon its brown ground than in the rug-work of the looms. How I loved to sit upon such rugs, when a child, and gaze at the strange figures, as they were lit up by the flashing fire-light! Memory recalled one very impracticable reindeer, which used to lie just in front of a maiden aunt's chair, representing a Brussels manufacturer's idea of the animal. His horns were longer than his head, body and tail combined, and the spring he was making, when transfixed by the loom, brought his nose so close to the ground, that my older boyhood calculated the immense antlers would certainly have tipped him over had he not been held back by the threads.

But to return to the plains. We examined high-lands and lowlands for poor soil, but found none. What we had once expected to see a bed of sand, if ever we saw it at all, turned up under the spade a rich dark loam, in depth and character fully equal to an Illinois prairie. Together with those other legends, localized drought and grasshoppers, the American desert, when revealed by the head-light of civilization, had taken to itself the wings of a myth, and fled away. There was a great sameness in the climate, as well as the scenery. Day followed day, with its sunshine and its winds, the latter being decidedly the most disagreeable feature of the entire trip.

Various episodes marked our journey from Silver Creek to Sheridan. A few only of the more noteworthy incidents can be transferred to these pages. They will suffice, however, as specimens of our adventures, and help the reader, I trust, to a better acquaintance with the free, wild life of the West.

The second day after leaving Silver Creek, we suddenly encountered another specialty of the plains, the "Wild Huntress." So often has this personage and her male counterpart danced, with big letters and a bowie-knife, across yellow covers, that we met the "original Jacobs" of the tribe gleefully. She came to us in a cloud of buffalo, with black eyes glittering like a snake's, and coarse and uncombed hair that tangled itself in the wind, and streamed and twisted behind her like writhing vipers. A black riding habit flowed out in the strong breeze, its train snapping like a loose sail, and a black mustang fled from her Indian lash—the dark wild horse, a fit carrier for such somber outfit.

She was introduced to us by the bison herd, which came thundering across our front, with this strange figure pressing its flank and darting hither and thither from one outskirt of the flying multitude to the other. The reins lay loose on the neck of her mustang, which entered into the fierce chase like a bloodhound, doubling and twisting on its course with an agility that was wonderful.

One hand of the huntress held out a holster revolver, which she fired occasionally, but with uncertain aim, one of the bullets indeed whistling our way. The chase constituted the excitement that she sought, and the pistol was little more than a spur to urge it on.

"That 's Ann, poor P—'s wife," said our guide. "Crazy since the Indians killed her husband. He was a contractor on the railroad; his camp used to be just above Hays. She lives in the old "dug-out" on the line yet, and spends half her time chasing buffalo. She never kills none, but that is n't what she is after. She wants to be moving, and just as wild as she can; it sort o' relieves her mind."

The huntress had seen our outfit, and rode toward us. The face was a very plain one, with a vacant yet anxious expression, and the tightly-drawn skin seeming scarcely to cover the jaw-bones. She halted before us, and commenced conversation at once.

"Good day, gentlemen."

"Good day, madam."

"She always tells her story to every body," muttered the guide in a low voice.

"Have you seen any Cheyennes hereabouts, gentlemen? I sighted a party this morning, and you ought to have seen them run. Raven Dick, here, put his best foot foremost, but they shook him out of sight in a ravine. Have n't any thing better to do, friends, and so I 'm riding down some buffalo."

We could easily understand why superstitious savages should run when a maniac female of such dismal aspect flitted along their trail.

"Out from Hays, sirs?" she continued, after a

pause. "I left there yesterday. Dick and I camped last night. We must be home when the men come in from work this eve. Up, Rave!" and she struck the mustang a cruel blow, from which he jumped with quivering muscles, only to be violently curbed. For the first time she had just noticed our guide, and sat for an instant with her wild eyes eating a way to his heart. Then she turned again to us.

"Sirs, you must aid me. Some say the Cheyennes killed my husband, and others there be who think Abe there did it. More shame to me who has to tell it, but the two had a fight about a woman, some months gone. It was just after pay-day, and husband was drunk; otherwise he 'd never have bothered his head about any girl but the one he married.

"There were blows and black eyes, and being a rough man's quarrel, it ended with hand-shaking. My man came home, and we sat by the fire that night, and I took no notice that he 'd been wrong, but spoke of our old home in Ohio, and asked him would n't he go back there when the contract was finished. And he put his hand on mine, and says: 'Sis, if the cuts and fills on the next mile work to profit, we 'll go home.' Just then there came a hiss from the door at our backs, and husband turned sharp and quick. There was a knot-hole in the planks, and its round black mouth, gaping from out in the night at us, had spit the sound into our ears. Husband he rose and went to the door, and fell back dying, with an arrow in his breast. Some said it was a Cheyenne, and others said Abe did it. There were lots of Indian bows in camp, and Cheyennes do n't kill for the love of it, but only to steal. I 'm going to ask them, if I can catch them, did they do it, and if not, I know who did. I 've a bow, Abe, and an arrow too, and I hope his blood is n't on your hands."

"I did n't do it, Ann. I do n't shoot no man in the dark," replied our hostler guide, with a sullen defiance, which among that class stands equally well for innocence or guilt. We looked at the two, as they sat for an instant facing each other. The picture was a weird one—a wildcat, fronting the object of its chase, but undecided whether to spring or not. We felt that the dark maniac had been hovering around us, and that this meeting was not altogether accidental. Her disordered brain was yet undecided in which direction vengeance lay, and, like a tigress, she was watching and waiting.

Our policy developed, on the instant, into a non-committal and a safe one. As she wheeled her horse, and left us without a word, we remarked to our guide that crazy folks were often suspicious of their best friends.

"That 's so," he replied, and rode off to urge on the wagons. We shrank from the idea of living with a murderer, and acquitted him of the crime on the spot.

We are moving out over the grand, illimitable plain again. Reader, ride with us awhile by the side of that big bison bull, which we have just stirred up from his noonday dream. You see his broad nostrils, reddish just under the dark skin at the end, and sensitive as the nose of a pointer. They have caught the air which we tainted, while passing for a moment across the breeze.

ONE OF OUR SPECIMENS.

BUREAU OF ILLUSTRATION BUFFALO. N.Y.

He has seen nothing, and we are still invisible, but he does not stop to look behind. "Escape for your life!" has been as plainly telegraphed from nose to brain, as it could be by eyes or mouth. We were so far off and well hidden then, that those active tell-tales, sound and sight, could play no part in this alarm. But the sentinel nerves of smell fled back from their post on the frontier, with the cry of "Man!" and the beast of the wilderness thinks only of flight. Powerful for defense against the rest of the animal creation, he is coward on the instant before its king.

Away he goes, right into the teeth of the wind, which he knows will tell him of any other foes ahead. Lumber along, old fellow, in your ponderous gallop,—the reader and I are on your path. Our saddle girths have been tightly drawn, the holster pistols are nestled snug at hand, in their cases on either side of the saddle-horn, while across its front lies the light Henry carbine, with a shoulder-strap attaching it to our person, should we drop the gun for the pistol. Thus we ride with twenty-four shots before reloading, at the service of our trigger-finger; the carbine carries twelve, the pistols each a half-dozen.

How warm we have become. Our hearts are as high up as they can get, bumping away at the throat-valves, as if they wished to get out and see what it is that has called their reserves into action.

There is a muskish taint in the air, from the game ahead. Put in your spurs, comrade; do n't spare. Get up beside him quickly as possible. Once there, the horses will easily stick. A stern chase disheartens the pursuer, encourages the pursued. Look out for that creek! See how the buffalo takes its steep bank—a plunge headlong, which sends the dust up in clouds. Now, as we check and turn into a ford, he is going up the opposite side.

Another hundred yards, and we are close beside him. The long tongue is hung out, and his head lies low down, as he plunges steadily forward, diverging ever so little as we press up opposite his fore-shoulders. That was a bad shot, my friend, barely missing your horse's head. Shooting at full gallop is like drawing straight lines while being shaken.

Some of our bullets are telling; you can hear them crack on his hide. There is a red spot now, not bigger than the point of one's finger, opposite a lung, and drops of blood trickle, with the saliva, from his jaws. Half a score of balls have been pelted into his big body, and he is bleeding internally. Now the blood comes thicker, and little clots of it drop down. He slows up—there is danger; look well to your seat!

That was a narrow escape, comrade. The bull suddenly whirled on his forefeet for a pivot, and your horse's chest, which was brushing his hind-quarters, grazed the black horns as they dipped for a plunge. The pony's swerve barely saved you both.

Now he stands sullen, glaring at us. The wounds look like little points of red paint, put deftly on his shaggy hide. They bleed inwardly, just crimsoning the brown hair at their mouths. The large eyes roll and swell with pain and fury. He is measuring our distance.

See him blow the blood from his nostrils. The

drops scatter like red-hot shot around him, seeming to hiss in globules of fury, as they spatter upon the dry grass. Bladder-like bubbles sputter in ebb and flow, from the red holes over his lungs. Tiny doors, for death's messengers to have entered in at.

What a marvel of size and ferocity he looks. Only our horse's legs stand between us and disembowelment. Down drops the head into battery again, and his rush would knock us over like nine-pins, did we stay to receive it. But bison charges are short ones. Our animals spring away, and he stops. Signs of grogginess are coming on him. How he hates to feel his knees shake, straightening them out with a jerk, as we thought he was just going down.

But at last gradually and gracefully he sinks, doubling his legs under him, and resting on his belly. There is still no flurry, or motion of any kind denoting pain. Unconquerable to the death, he suddenly falls on his side, the limbs stiffen, and he is dead.

Twine your hands in the long beard, and in the mane. How he shames the lion, for whom he could furnish coats half a dozen times over. What switches of hair those black fetlocks would make. Was there ever another so big a bison?

Wondering over this, we lie down on the prostrate bulk, and wait for the wagon.

CHAPTER XXI.

"CREASING" WILD HORSES—MUGGS DISAPPOINTED—A FEAT FOR FICTION—HORSE AND MONKEY—HOOF WISDOM FOR TURFMEN—PROSPECTIVE CLIMATIC CHANGES ON THE PLAINS—THE QUESTION OF SPONTANEOUS GENERATION—WANTON SLAUGHTER OF BUFFALO—AMOUNT OF ROBES AND MEAT ANNUALLY WASTED—A STRANGE HABIT OF THE BISON—NUMEROUS BILLS—THE "SNEAK THIEF" OF THE PLAINS.

WHILE we were at breakfast one morning, the guide ran in to say that the herd of wild horses which we had seen on Silver Creek, were feeding toward us, a mile away. I left the table to obtain a view of them, and by Abe's advice carried my rifle, as he suggested that we might "crease" one of them. This feat consists in hitting the upper edge of the bones of the neck with a bullet, the blow striking so high up that it will momentarily paralyze, without fracturing. We had read of it often in tales of Western daring, where the hero mounted the prostrate steed, and, upon its return to consciousness, escaped on its back from numberless difficulties and hosts of Indians.

A short distance out from camp, we turned and saw Muggs following us with a saddle and bridle on his arm. He had suffered grievous wrong at the heels of his mule, and was bent on possessing himself of one of our creased horses. After creeping, with almost infinite caution, within seventy-five yards, we succeeded in placing our bullets exactly where we intended, thereby knocking down two victims, who at once became insensible—and no wonder, for their bones were as effectually fractured as if they had been struck with a sledge-hammer. Muggs' faith in the theory of creasing, however, was unbounded. Up he ran and buckled on the saddle, and got one foot in the stirrup, ready to swing himself into the seat, when the animal rose.

After waiting about ten minutes, our Briton concluded that a dead horse was poor riding, and left us with a very emphatic statement that, in his opinion, capturing a mount with a rifle was "another blarsted Hamerican lie, you know!"

I afterward conversed with several plainsmen about the merits of "creasing," and found that their attempts had invariably ended in the same way as ours had done. The feat may have been possible with smooth-bore rifles, in the hands of those remarkable hunters of old, who were able to shoot away the breath of a pigeon, and hit the eye of a flying hawk; but with breech-loaders I unhesitatingly pronounce creasing an utter impossibility. The achievement sounds well in theory, but, like much else of popular Western lore is somewhat impracticable when fairly tested. I have an idea that the principal market value of "creased" horses in the future, as in the past, will be derived from furnishing creatures of romance with fearful rides. For this purpose, a cracked skeleton would be as apt as a sound one, to carry the rider into many of the scenes with which these tales are wont to harrow our souls.

While crawling up on the herd, we took its census very carefully. I was a little surprised to find there were but twenty-five horses, all told. They were apparently a little larger than the wild ones of Texas, and had bushy manes and tails, and their step was remarkably firm and elastic. They were exceedingly timid creatures, raising their heads constantly, to gaze around. One very interesting circumstance connected with the herd was that among these wild horses we noticed two strangers; one, a feeble old buffalo bull, expelled from his tribe, and seeking their aid against the wolves, and the other, the black pacing stallion.

When we fired, the survivors were off on the instant, and the manner in which their clean hoofs struck the earth, and spurned it, was truly worth seeing. No heaves either, it was plain to see, had ever troubled those full chests. We caught sight of the herd awhile after, on a ridge four miles away, and they were still running at full speed. These were the only wild horses we saw on our trip. In fact, but two or three small droves are believed to exist on the plains, as the great mass of the shaggy-maned thousands, children of those old Spanish castaways, swarm nearer the Pacific.

So timid and fleet are these horses that none of them have ever been captured except during the early spring. They are then poor, and, by hard spurring, can be ridden down. At other times their bottom, and the advantage of having no weight to carry, insure their safety. It is quite probable, however, that a systematic pursuit, of the kind practiced in Texas, might prove successful at any season of the year.

I gazed at our two victims with less satisfaction than at any thing I had ever killed. Shooting horses, dear reader, is a good deal like shooting monkeys. They are both too intimately associated with man to be made food for his powder. One is a very true and faithful servant, and the other, if we may believe Mr. Darwin, was once his ancestor.

In examining the two handsome bodies lying there, I noticed one fact to which I should have liked to draw the attention of the whole learned fraternity of blacksmiths, who mutilate horses, the world over. The hoofs were as solid and as sound as ivory, without a crack or wrong growth of any sort. And why? Turning them up, the secret lay exposed; for there, filling the cavity within—a sponge of life-giving oil—was the frog entire, just as Nature made and kept it. Its business was to feed and moisten the hoof, and this it had done perfectly. No blacksmith had ever gouged it out with his knife, and robbed it anew at every shoeing.

It is noticeable that the equine race, in its wild state, has none of the ills of the species domesticated. The sorrows of horse-flesh are the fruits of civilization. By the study and imitation of Nature's methods, we could greatly increase the usefulness of these valuable servants, and remove temptation from the paths of many men who lead blameless lives, except in the single matter of horse-trades. It may well be queried, perhaps, whether even the patient man of Uz, had he been laid up by a runaway colt instead of boils, could have resisted the temptation to trade it off upon Bildad the Shuhite, when that individual came to condole with him.

As we journeyed onward, we found the soil ever the same, in depth and strength equal to an Illinois prairie. The old cretaceous ocean, and the great lakes, certainly left it rich in deposits. When its surface shall have been broken by the plow, and the water-fall absorbed instead of shed off, the plains will resemble, in appearance and products, any other prairie country. The amount of moisture annually passing over them, in storm-clouds that burst further east, is abundantly sufficient to make the tract very fertile. It is a well established fact in relation to climatic influences, that moisture attracts moisture; and in this region the dry ground, with its few shallow streams, has now no claim upon the summer clouds. The tough buffalo grass has put a lock-jaw on the plain. It can drink nothing from the floods of the rainy season. But pry open the hungry mouth with the plowshare, and the earth will drink greedily. The moisture then absorbed, given up through the agency of capillary attraction, will draw the showers of summer, as they are passing over. Already a marked change has taken place over a portion of the plains, and crops have been grown as far west as Fort Wallace.

The subject of spontaneous generation, I may remark in this connection, became a very interesting one to our party. Wherever the soil has been dis turbed, wild sun-flowers spring suddenly into existence. The "grading camps" of the railroads were followed by belts of these self-asserting annuals. The first garden-patch cultivated at Fort Wallace had weeds and insects similar to those that infest gardens elsewhere. In some cases hundreds of miles of barren plain intervened between the spots where the seeds germinated, and the nearest points where other plants of the same variety grew. Neither birds or wind could have carried the seeds in such quantities. Is the theory true that germs fall down to us from other planets? Or, do not the plains offer a strong argument on behalf of spontaneous generation?

Another matter on which the plains appealed to us strongly, pertained to the wanton destruction of its wild cattle. During the year 1871, about fifty thousand buffalo were killed on the plains of Kansas and Colorado alone. Of this number, it will be correct to estimate that about one-third were shot for their robes, as many more for meat, and sixteen thousand or so for sport. Each buffalo could probably have furnished five hundred pounds of meat and tallow, the quantity of the latter being small. When killed for food, only the hind quarters and a small portion of the loin are saved, in all perhaps two hundred pounds. The hides of these are sacrificed, the skin being cut with the quarters, and left on them for their protection. The profits of this great slaughter would, therefore, be about 16,500 robes and 3,300,000 pounds of meat; the waste over 33,000 robes, and probably not less than 20,000,000 pounds of meat.

In this computation, the vast herds which range further north are not included. There, however, the waste is comparatively small, as the red man is in the habit of saving the greater portion of the flesh and robes. Of the above twenty million pounds of meat left to rot in the sun, and taint the air of the plains, the greater proportion would furnish sweeter and more nourishing food to the poor classes of our cities than the beef which they are able to obtain.

Let this slaughter continue for ten years, and the bison of the American continent will become extinct. The number of valuable robes and pounds of meat which would thus be lost to us and posterity, will run too far into the millions to be easily calculated. All over the plains, lying in disgusting masses of putrefaction along valley and hill, are strewn immense carcasses of wantonly slain buffalo. They line the Kansas Pacific Railroad for two hundred miles.

Following ordinary sporting parties for an hour after they have commenced smiting the borders of a herd, stop by a few of the monsters that they leave behind, in pools of blood, upon the grass; draw your hunting-knife across the fat hind-quarters, and see how the cuts reveal depths of sweet, nourishing meat, sufficient to supply two hundred starving wretches with an abundant dinner; then if your humanity does not tempt to a shot at the worse than pot-hunters in front, God's bounties have indeed been thrown away upon you.

By law, as stringent in its provisions as possible, no man should be suffered to pull trigger on a buffalo, unless he will make practical use of the robe and the meat. What would be thought of a hunter, in any of the Western States, who shot quails and chickens and left them where they fell? Every citizen, whether sportsman or not, would join in outcry against him. Another matter which the law should regulate relates to the protection of the buffalo cows until after the season when they have brought forth their young. The calf will thrive, though weaned by necessity at a very early age, and the season for shooting cows, although short, would be amply long enough to comport with the chances of future increase.

Probably the most cruel of all bison-shooting pastime, is that of firing from the cars. During certain periods in the spring and fall, when the large herds are crossing the Kansas Pacific Railroad, the trains run for a hundred miles or more among countless thousands of the shaggy monarchs of the plains. The bison has a strange and entirely unaccountable instinct or habit which leads it to attempt crossing in front of any moving object near it. It frequently happened, in the time of the old stages, that the driver had to rein up his horses until the herd which he had startled had crossed the road ahead of him. To accomplish this feat, if the object of their fright was moving rapidly, the animals would often run for miles.

When the iron-horse comes rushing into their solitudes, and snorting out his fierce alarms, the herds, though perhaps a mile away from his path, will lift their heads and gaze intently for a few moments toward the object thus approaching them with a roar which causes the earth to tremble, and enveloped in a white cloud that streams further and higher than the dust of the old stage-coach ever did; and then, having determined its course, instead of fleeing back to the distant valleys, away they go, charging across the ridge over which the iron rails lie, apparently determined to cross in front of the locomotive at all hazards. The rate per mile of passenger trains is slow upon the plains, and hence it often happens that the cars and buffalo will be side by side for a mile or two, the brutes abandoning the effort to cross only when their foe has merged entirely ahead. During these races the car-windows are opened, and numerous breech-loaders fling hundreds of bullets among the densely crowded and flying masses. Many of the poor animals fall, and more go off to die in the ravines. The train speeds on, and the scene is repeated every few miles until Buffalo Land is passed.

Another method of wanton slaughter is the stalking of the herds by men carrying needle-guns. These throw a ball double the weight of the ordinary carbine, and the shot is effective at six hundred yards. Concealed in ravines, the hunter causes terrible havoc with such weapons before the herd takes flight. We were never guilty of ambushing after those two days on the Saline, and of those occasions we were heartily ashamed ever afterward.

One specialty of the plains that deserves mention, and quite as remarkable as its brutes and plants, though of rather more modern origin, is its numerous Bills. Of these, we became acquainted, before our trip was ended, with the following distinct specimens: Wild Bill, Buffalo Bill, California Bill,

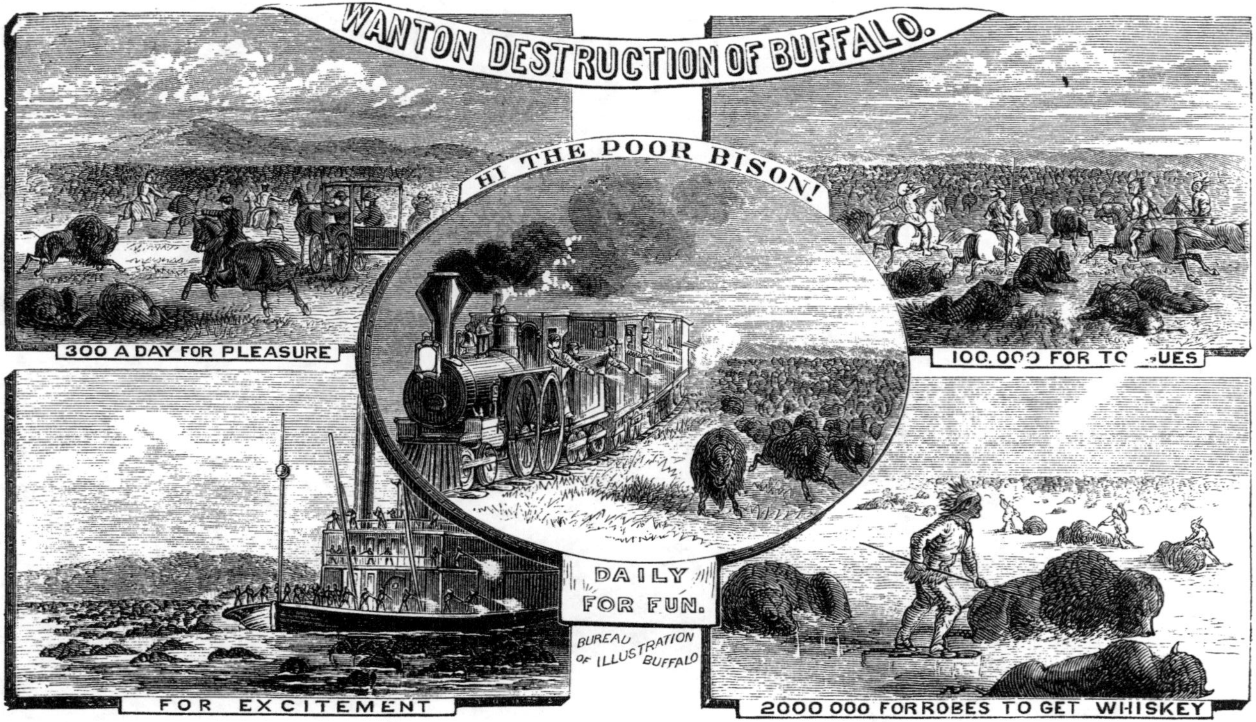

WANTON DESTRUCTION OF BUFFALO.

HI THE POOR BISON!

300 A DAY FOR PLEASURE

100.000 FOR TO UES

FOR EXCITEMENT

DAILY FOR FUN.

BUREAU OF ILLUSTRATION BUFFALO

2000 000 FOR ROBES TO GET WHISKEY

Five pictures for the consideration of Uncle Samuel, suggestive of a game law to protect his comb-horns, buttons, tallow, dried beef, tongues, robes, ivory-black, bone-dust, hair, hides, etc.

Rattlesnake Bill, and Tiger Bill, the last named being, as one of our men who had played with him remarked, the "dangererest on 'em all." We also heard of a Camanche Bill and an Apache Bill, but these celebrities it was not our fortune to meet.

I can not dismiss the peculiar characters of the plains without again paying tribute to that unapproachable thief, the cayote. Let no party of travelers leave any thing exposed in camp lighter than an anvil. We lost, in one night, at the hands—or rather the jaws—of these slinking sneak-thieves of the plains, a boot, a pair of leather breeches, and a half-quarter of buffalo calf, besides some smaller articles.

CHAPTER XXII.

A LIVE TOWN AND ITS GRAVE-YARD—HONEST ROMBEAUX IN TROUBLE—JUDGE LYNCH HOLDS COURT—MARIE AND THE VINE-COVERED COTTAGE—THE TERRIBLE FLOODS—DEATH IN CAMP AND IN THE DUG-OUT—WAS IT THE WATER WHICH DID IT?—DISCOVERY OF A HUGE FOSSIL—THE MOSASAURUS OF THE CRETACEOUS SEA—A GLIMPSE OF THE REPTILIAN AGE—REMINISCENCES OF ALLIGATOR-SHOOTING—THEY SUGGEST A THEORY.

OUR fourth day's travel from Silver Creek brought us to Sheridan, our secondary base of operations, so to speak, and only fourteen miles east of the Colorado border. We found the town a very lively one, notwithstanding that the grave-yard, beautifully located in a commanding position overlooking the principal street, was patronized to a remarkable extent. The place had built itself up as simply the temporary terminus of the Pacific Railroad. Soon after our visit it moved westward, and at last accounts but one house remained to mark its former site.

The shades of night had just settled over the town upon the evening of our arrival, when Abe, our hostler-guide, came running to us with information that "Honest Rombeaux," another of our hostlers, was being hung by some of the citizens. The locality which had been selected for this little diversion was a railroad trestle a short distance below the town. We were already acquainted with the penchant our Sheridanites had for hanging people. Thirty or more graves on the neighboring hill had been pointed out before sundown, as those of persons who had fallen under sentence from Judge Lynch. In the expressive language of the citizen who volunteered the information, there had been "thirty funerals, and not one nateral death." Now that Judge Lynch had opened court at our own door, we proposed to raise the question of jurisdiction.

Armed, at once, we set off for a rescue, and, stumbling through the darkness, had gone only a hundred yards or so, when we met the lynchers returning. At their head, with a very dirty piece of rope around his neck, walked our hostler, trembling all over, and chattering broken English rapidly, in mingled fright and anger. The leader of the party told us that the evidence not being quite sufficient for hanging, an extra session of court had been called to be held immediately, and as having some interest in the case, we were invited to seats on the jury. The trial, we were further informed, was to be held in Rombeaux's own house. This last was a new surprise, for reasons to be explained presently. Rombeaux had been with us ever since leaving Hays, and had gained his title of "Honest" from a particularly faithful discharge of duty.

To him had been intrusted the supplies for hired men and horses. Three of the Mexicans he had severally thrashed for stealing. Once, in the night, on Silver Creek, we had heard a rattling at the medicine-chest, and trembling for our limited stock of spirits, stole forth to catch the culprit. On his knees by the open box was Rombeaux, replacing the brandy-bottle, and we feared that he, too, had become a thief. But just then, on the still air, came words of thanks to the Virgin Mary, for having enabled him to awake in time to frighten away the robber. Nor was this all; in the fierceness of his indignation, we beheld him sally forth immediately afterward, and kick a sleeping Mexican out of his blankets, on suspicion. Thereupon, we went back to bed with implicit faith in Rombeaux, which had followed us ever since.

Had he not told us, moreover, of a vine-covered cottage in France, where pretty Marie watched and waited until her lover could earn dowry sufficient to match hers? It was the old story. A maiden fair tarried in Europe, while a true knight ransacked foreign lands for fame and fortune; and long since had all of us, save Sachem, exhausted our stock of spare change to hasten the reunion.

Passing some of the lowest and most flashy-looking saloons in the place, we entered a ravine, and soon stopped before a "dug-out." So much was it the work of excavation, that the dirt roof was level with the earth above, and the door seemed to open directly into the bank. We knocked, and were answered promptly by a fat, gayly dressed French woman. This was Rombeaux's wife, and here was Rombeaux's house. What a Marie and vine-clad cottage these!

Without delay the trial commenced, the Frenchman and his wife occupying places in the center, and the court seated on boxes, barrels, and the bed. The evidence taken that night in the cabin was substantially the following:

Two years before Jules Pigget, a native of France, accompanied by his young wife, appeared on the railroad below, and solicited work. They both found ready employment, and lived below Hays, in a dugout, happy and prosperous. Within a year came another Frenchman, our present Honest Rombeaux. Across the water, he and Jules had been rival suitors for Marie's hand; yet strangely enough, the newcomer was welcomed by the young couple, and took up his abode with them. Matters prospered with all three, and soon Jules was to be appointed tank-tender on the road. That year came the great rainstorm, when so many families in Western Kansas and Texas were drowned. Hundreds of people were living in dug-outs, rude excavations in the banks of streams, with the roof on a level with the bank above, but the room itself entirely below high-water mark—a style of dwelling which, as no great rise had occurred in years, had become quite popular among new-comers.

On the night of the great flood people went to bed as usual. The streams had risen but little. At midnight the rain fell heavily; the firm surface of the plains shed the waters like a roof; streams rose ten feet in an hour, and the foaming currents, roaring like cataracts, came down with the force of mighty tidal waves. Many dwellers in the dug-outs sprang from their beds into water, to find egress by the doors impossible, and were fortunate if they succeeded in escaping through the chimneys or roofs. Whole families were drowned. Fort Hays, at the fork of Big Creek, and supposed to be above high-water, was inundated, six or eight soldiers being swept away, while the remainder were obliged to seek safety on the roofs of the stone barracks. Large numbers of mules, picketed on the adjacent bottoms, were drowned. Their picket-pins fast in the earth, the animals were swept from their feet by the rising waters, and towed under by the firmly-held lariats. Emigrants encamped on the bottom heard the roar of the flood; with no time to harness, they seized the tongues of their wagons themselves, but the rising tide gained on them too rapidly, and they were glad to save life at the expense of oxen and goods. The horrors of that night are indescribable, and, to crown all, they took place amid a darkness that was total. Above, was the roar of waters descending; below, the answering roar of the floods, as they rolled madly onward, carrying in their strong arms the wreck of farms, and corpses by the score.

On that night Jules, the husband, perished. Honest Rombeaux and Marie, however, were rescued from the roof of their dwelling at daylight; and afterward, when the flood had subsided, the body of Jules was taken from the wash in the fire-place. And now came suspicion, and pointed over the shoulders of the throng gathered around; for there was an ugly wound half hidden in the dead husband's hair, and his fingers were bruised. Some men did not hesitate to say boldly that when Rombeaux escaped through the chimney, Jules stayed behind to assist his wife out, and that when he tried to follow, he was struck on the head by his quondam rival, and, still clinging to the chimney's edge, his fingers were pounded until their hold was loosed, and the victim sucked under the roof, against which the waters were already beating. The man and woman, however, claimed that it was the whirl of the waters against pegs and logs which had disfigured the corpse. Three weeks afterward they were married.

"And now, gentlemen," said our foreman, rising from his barrel, when the evidence was all in, "the question for the jury to decide is, Was it the water that did it?"

A doubt existing in the case, we gave the prisoner its benefit; but there was murder in the air, and Rombeaux knew it. Before morning he had departed—Marie said for La Belle France, but, as the citizens generally believed, really for Texas.

The next twenty-four hours constituted a regular field-day for the Professor, being distinguished by an event which, from a scientific stand-point, was among the most important of our entire expedition. This was the discovery of a large fossil saurian, which we came upon while exploring quite in sight of Sheridan, and not more than half a mile from its eastern outskirts.

Descending the side of a deep, desolate rift in the earth, we found ourselves among unmistakable traces of violent volcanic action. The ground was strewn with black sand, and with yellow pebble-like masses, apparently impure sulphur. There were numerous round cones also, looking like diminutive craters, with edges and surface composed of bubble-like lava, the material having evidently hardened while still distended by the struggling gases. The appearance, to use a homely comparison, was somewhat that of several low pots, over the edges of which boiling molasses had poured, and then burned by the heat of the fire. Some scattered objects, which at first we took for stumps of huge trees, upon examination we found to be pillars of mud and rock, upheavals, apparently, from volcanic action, and not the work of the floods, which, in those primeval times, we knew, must have poured down the valley. They would have answered, without much difficulty, for druidical altars, had we only been in the land once inhabited by those long-bearded, blood-thirsty priests of old.

Two or three poisoned cayotes and a dead raven were lying near some bleached buffalo skulls, on which, as we presently discovered, daubs of lard mixed with strychnine had been placed, and licked off by the victims; and straightway, as genius of the scene, an unshaven, woolen-shirted little man appeared in view, busily engaged in skinning a wolf. We saluted him, and the response in French-English told us his nationality at once. We found his name to be Louis, and his proper occupation that of watchmaker. But as the pinchbeck time-pieces of the frontier did not furnish enough repairing to take up his entire time, he had many spare hours, and these he devoted to securing pelts. As buffalo were not now in the vicinity, he larded their bones, with the success of which we were eye-witnesses.

Louis was a wiry little Gaul, very positive in his

ideas about every thing. An animated conversation sprang up at once between him and the Professor, and it soon became amusingly evident that his geological ideas did not entirely accord with those of the Philosopher. A sudden turn in the colloquy developed a fact of keen interest to even the most unscientific member of our party.

Pointing to the other side of the valley, Louis told us that there lay the bones of an immense snake, all turned to stone. This sudden voice from the past ages sounded in the Professor's ears like the blare of a trumpet to a warrior. He hurried us forward in the direction indicated, and, locking arms with the bloody-shirted little Frenchman, strode on in advance. I wish his class could have seen him thus traversing the desolate bed where that old sunken volcano went to sleep. We were glad that the latter was still asleep, and had never acquired the habit of snorting into wakefulness, and pelting explorers with hot rocks.

What mysteries, I have often thought, might we not discover, on looking down the throat of a healthy volcano, if some wise alchemist could only brew a dose sufficiently powerful to stop the fiery fellow's foaming at the mouth! Or, better still, if it could reach the bowels of the earth, and keep the whole system quiet, while we, puny mortals, like trichina mites, swarmed down the interior, and bored scientifically back to the crust again. Earth's veins run golden blood, and we might be gorged with that, perhaps, ere making exit into the sunshine again.

A shout from the further edge of the ravine cut

short our speculations, and called our attention to the Professor. He stood waving his slouched hat for an instant, and then bent close over the ground, in earnest scrutiny.

A few moments later, and we all stood beside the huge fossil. It lay exposed, upon a bed of slate, looking very much like a seventy-foot serpent, carved in stone. Part of the remains had been taken up to the town, and spread over the bench, in the shop of Louis. From what was left, the jaws appeared to have been originally over six feet long, the sharp hooked and cone-shaped teeth being still very perfect. A few broad fragments of ribs showed that, in circumference, the animal's body had been about the size of a puncheon. We felt confident that the specimen was a very rare one, as Muggs had never seen any thing like it, even in England. It now rests in the museum at Cambridge, Massachusetts.

"This fossil, gentlemen," said the Professor, "is that of a *Mosasaurus*, a huge reptile which existed in the cretaceous sea. This appears to be one of the largest members of the family yet discovered, its length, as you will perceive, being over fifty feet. The species to which it belonged swarmed in immense numbers, but were surrounded by monsters even more remarkable than they. The deep which they inhabited must have been constantly lashed and torn with their fierce conflicts; for it was an age of war, and the powers of offense and defense, which the monsters of that period possessed, were terrible. Winged reptiles filled the air, in appearance more hideous than any creation of the imagination. Following close upon the Reptilian came the Mammalian age, and I hold that with the largest of the mammals came man, rude in tastes and uncouth in form, but even then ruling as king of the animal creation. Wielded by a strength equal to that of a gorilla, his club would dash in the skull of any beast which dare dispute dominion with him."

The text thus suggested him, the Professor then diverged into an argument on his pet theory of man's early existence.

A trivial circumstance connected with our discovery arrested my attention, and, from a sportsman's stand-point, suggested a little theory of my own. The head of the saurian rested on the basin's edge, its jaws touching, with their stony tips, the prairie, while down into the valley below stretched the body and tail. This little fact dove-tailed itself into some incidents of the past, and gave rise to quite a train of speculation.

Some years ago I hunted alligators in Mississippi. Sitting on the bank of a sluggish bayou, I would watch the surface of the water, close under which were visible the noses of countless buffalo fish, floating as one sees minnows do in glass jars. Under the hot sun all nature seemed asleep. Soon, however, a black knot, an ugly dark wart, not larger than one's two fists, would make its appearance, floating, like some charred fragment, slowly along.

To a stranger, the only suspicious circumstance would have been, that where there was no current whatever, it still continued its motion, the same as before. The experienced eye recognized this object as the nose of an alligator, behind which, and just at the surface, as it got opposite, the ugly eyes would become visible, looking out for hogs or dogs, as they came to drink under the bank.

I never had the patience to wait for the *finale* of the scene; but had I done so, I should have beheld the knot float closer in, and, just after passing the victim, a tail would have come out of the water, and, with a curving blow forward, knocked the prize out from shore, and in front of the devourer's jaws. It was my good fortune, frequently, to send a Ballard rifle-ball into the pirate's eyes. In such cases there was usually a tremendous commotion in the water, accompanied by a strong smell of musk, and the wounded reptile would then make straight for shore, and run his head upon it. Under such circumstances, the creature always sought at least that much of dry land to die upon, seeming as anxious as man that its lamp of life should not be extinguished under water.

This monster whose remains we were now exhuming was allied to the alligator, as one of the great family of lizards, and had died in the same manner—his head on the shores of the basin, his tail in its depths. Perhaps in the convulsion of Nature which opened a path for the waters to the ocean, and drained this inland sea, the fissure in which we stood had gaped, and exhaled poisonous gases through the whirlpool its suction created. The saurian monster of that strange age felt the hungry vortex swallowing him, which meanwhile enveloped him in deadly secretions, killing before devouring. With a last lurch through the cauldron's ebbing tide, the lizard threw himself upon its edge, and died.

Of the countless millions of saurians then existing, capricious Nature had seized upon this one, to transmute it into an imperishable monument of that extinct race. In those ages of roaring waters and hissing fires, she had clothed the bones in stone, that they might withstand the gnawing tooth of time, and thus handed them down to the wondering eyes of the Nineteenth Century. Many of the pieces, it should be said, were cracked and scarred, evidently by the action of fierce heat.

Constantly the earth is giving up these marvelous creations of the past, in comparison with which the animals of the present are tame enough. While we doubt a modern sea-serpent as impossible, we dig up fossilized marine monsters, which could easily have swallowed the biggest snake that credible sea-captain ever ran foul of.

DUG-OUT.

CHAPTER XXIII.

FROM SHERIDAN TO THE ROCKY MOUNTAINS—THE COLORADO PORTION OF THE PLAINS—THE GIANT PINES—ATTEMPT TO PHOTOGRAPH A BUFFALO—THINGS GET MIXED—THE LEVIATHAN AT HOME—A CHAT WITH PROFESSOR COPE—TWENTY-SIX INCH OYSTERS—REPTILES AND FISHES OF THE CRETACEOUS SEA.

AT Sheridan, we were very near the Colorado portion of the plain, which stretched on for some hundreds of miles further westward, its further line lapping the base of the Rocky Mountains. Into this territory we passed, and spent a considerable period of time in its examination, but while our experience was to us full of interest, any thing more extended than a brief summary would occupy too much space here.

For the first one hundred miles, the soil deteriorated in quality, and the sage-bush made its appearance, as did also the "Adam's needle" or "Spanish bayonet." The latter makes an excellent substitute for soup, but a wretched cushion to alight upon when thrown from your horse. (I make the latter statement on the authority of Doctor Pythagoras.) Brackish water was found at intervals, and white saline crystallizations were seen along some of the streams. Although the soil was more sandy than further east, the buffalo grass was abundant and nutritious, so that at no time had we any difficulty in finding grazing for our cattle, and the antelope that we killed were invariably in good condition. This belt of eastern Colorado proved particularly rich in fossil wealth, to the description of which we shall devote most of this chapter, and the whole of that following. In the vicinity of the Big Sandy, we found numerous lakes of clear water, surrounded by rich pasturage.

About one hundred miles west of the Kansas line, the country began gradually improving, and continued to do so until we reached the mountains. The Bijou basin, through which we passed, afforded excellent range, and contained good streams. The country swarmed with antelopes, and once we saw a herd running rapidly, which was four minutes in crossing the road.

We had fine views of Pike's Peak, at a distance of one hundred and fifty miles, the atmosphere there being remarkably pure and transparent. Emigrants have often been deceived when, as their wagons crawled over the crest which we named First View, the fine old Peak burst upon their sight, and in their enthusiasm resolved to get an early start next day and reach it before another night-fall. Our guide told us that when he first crossed the plains, by the Platte route, his party camped for the night near Monument Rock. After supper, two of the men and a woman set out to cut their names in the stone, supposing it to be only a mile or so distant, but when an hour's traveling brought the rock apparently no nearer, they became discouraged and returned. Next day Monument Rock was found to be twelve miles distant from their camping-place.

When within a day's journey of the mountains, we came in sight of several tall objects standing out in bold relief upon the plain. These proved to be giant pines, thrown out, like sentinels, from the forests still far beyond and invisible. We could not resist the impulse to give the first one we came to a hearty hug; for, after so many weeks upon the treeless plain, these suggestions of mighty forests, with their mingled sheen and shadow, were indeed welcome. The mountains of Colorado, with their beautiful parks and wonderful young cities, have been so often described that our notes would prove a useless addition to a somewhat worn history, and hence we forbear taxing the reader's patience by transcribing them here.

After studying the principles of mining and irrigation, we spent in the neighborhood of one calendar month in getting views of sunrise and sunset, from all the known peaks, to the end that no future tourist might feel called upon to extend to us his kind commiseration for having lost some particular outlook, where he had been, and which he considered the best of all. To accomplish this thoroughly, we hewed paths up hitherto inaccessible mountains, and at the end of the month made a close calculation, and decided that we were a match for all such tourists for at least five years to come. We then retraced our steps to Buffalo Land, again entering the fossil belt near Fort Wallace.

One incident of our trip into Colorado deserves especial mention from having been the first, as it will probably prove the last, attempt to photograph the buffalo in his native wildness, at close quarters. The idea was suggested in a letter which the Professor received from his Eastern friends, who thought that actual photographs of the animals inhabiting the plains would be a valuable addition to the ordinary facilities for the study of natural history. As good fortune would have it, there happened to be at Sheridan an artist, just arrived from Hays, then prospecting for a location, and him we promptly engaged. The second day out, two old buffaloes, near our road, were selected as good subjects for first views. One of these was soon killed, the other making his escape up a ravine near by. Although we had good reason to suspect that the latter had been wounded, we did not pursue him, since it was now near noon, and our artist, moreover, being of a somewhat timid disposition, had expressly stipulated that we should keep near him, not so much, he repeatedly assured us, as a body-guard for himself, as for the protection of his new camera and outfit.

The dead bull we propped into position with our guns and other supports, and while the artist carefully adjusted his instrument, Shamus began to make preparations for lunch, and Mr. Colon and Semi set out for a few minutes' pastime in catching bugs. They had been gone a full half hour, and we were just remarking their prolonged absence somewhat impatiently, when a loud cry from the nearer bank of the ravine fell on our ears, and looking around we beheld Colon senior, and ditto junior, making toward us at a tremendous rate of speed.

"Buffalo!" was all that we could catch of Semi's wild shouts, as he led the chase directly toward us, his father having lost several seconds in securing one of his specimen-cases, and on the instant the old bull that we had wounded an hour before hove in sight, in full charge upon the flying entomologists. As buffalo charges are short ones, he would have stopped, no doubt, in a moment or so, had not Muggs and I, the only members of our party who happened to have their guns at hand, opened fire on him, and planted another bullet between his ribs. The effect was to infuriate the old fellow tenfold, and down he came careering toward us, with what I then thought the most vicious expression of countenance I had ever seen on a buffalo's physiognomy.

The attack was so sudden, and the surprise so complete, that we were most ingloriously stampeded, and fell back in hot haste upon our reserves, the guide and teamsters, who, we knew, would be provided with weapons and in good shape to cover our retreat. The sitting for which we had made such elaborate preparations was abruptly terminated in the manner shown in the accompanying engraving.

Fortunately for the artist, the blow originally intended for him was delivered upon the legs of the instrument. His assailant being at length dispatched, the poor fellow proceeded to pick out of the ruins of his property what remained that might again be useful. He stated that his stock, as well as the subject of buffalo photographing, was "rather mixed," and that, if we would pay him for the damage done, he would return. Next morning he left us, and thus it was that science lost the projected series of valuable photographic views.

TAKING AND BEING TAKEN.

Exploration gives us a past history of the plains which is interesting in the extreme. Our party spent some weeks in exploring for fossils beyond Sheridan, and were richly rewarded. In the great ocean which once covered the land, the wonderful reptiles of the cretaceous age swarmed in prodigious numbers, and their fierce struggles upon and under its surface made "the deep to boil like a pot." The mysterious Leviathan, described in the forty-first chapter of Job, had its prototype in more than one of the monsters of that period:

"Who can open the doors of his face? his teeth are terrible round about.

"Out of his mouth go burning lamps, and sparks of fire leap out.

"Out of his nostrils goeth smoke, as out of a seething pot or caldron.

"His breath kindleth coals, and a flame goeth out of his mouth.

"The flakes of his flesh are joined together: they are firm in themselves; they can not be moved.

"He esteemeth iron as straw, and brass as rotten wood.

"He maketh a path to shine after him; one would think the deep to be hoary."

The fossil remains of these reptiles are numerous, constituting a rich mine of scientific wealth, which has been but very lightly worked. Enough fossils can be obtained by future exploration to fill to overflowing all the museums of the land.

We have no means of computing how long the cretaceous sea existed, but we know that it passed away and was replaced by large fresh-water lakes, those of the plains being bounded on the west by the Rocky Mountains. Then succeeded an age of which we can catch but occasional glimpses, and our longing becomes intense that we could know more. We see a land fertile as the garden of Eden, surrounding beautiful lakes. The climate is delightful, and earth, air, and water, are full of life. Grand forests and flower-covered prairies nod and blossom under the kind caresses of Nature. Water fowls numberless plunge under and skim over the surface, and the songsters of the air warble forth their hymns of praise. Over the pastures and through the forests roam an animal multitude of which we can have but faint conception, but among the number we recognize the lion with his royal mane, and the tiger with his spots; and there also are the elephant, the mastodon, the rhinoceros, the wild horse, and the great elk.

After our return, the eminent naturalist, Prof. Edward D. Cope, A. M., visited the plains, and spent some time in careful exploration there. As he had previously received several fossils from us for examination, I communicated with him not long since, asking a record of his trip. This he very kindly consented to furnish, and, did space permit, I would gladly publish entire the matter which he has placed at my disposal. No apology can be necessary, however, for yielding to the temptation of devoting two or three chapters to a chat by Prof. Cope with my readers.

The manuscript, as it lies before me, is entitled: "On the Geology and Vertebrate Palæontology of the Cretaceous Strata of Kansas." Let us begin with "Part I—A General Sketch of the Ancient Life."

That vast level tract of our territory lying between Missouri and the Rocky Mountains represents a condition of the earth's surface which has preceded, in most instances, the mountainous or hilly type so prevalent elsewhere, and may be called, in so far, incompletely developed. It does not present the variety of conditions, either of surface for the support of a very varied life, or of opportunities for access to its interior treasures, so beneficial to a high civilization. It is, in fact, the old bed of seas and lakes, which has been so gradually elevated as to have suffered little disturbance. Consistently with its level surface, its soils have not been carried away by rain and flood, but rather cover it with a deep and widespread mantle. This is the great source of its wealth in Nature's creations of vegetable and animal life, and from it will be drawn the wealth of its future inhabitants. On this account its products have a character of uniformity; but viewed from the standpoint of the political philosopher, so long as peace and steam bind the natural sections of our country together, so long will the plains be an important element in a varied economy of continental extent.

But they are not entirely uninterrupted. The natural drainage has worn channels, and the streams flow below the general level. The ancient sea and lake deposits have neither been pressed into very hard rock beneath piles of later sediment, nor have they been roasted and crystallized by internal heat Although limestone rock, they easily yield to the action of water, and so the side drainage into the creeks and rivers has removed their high banks to from many rods to many miles from their original positions. In many cases these banks or bluffs have retained their original steepness, and have increased in elevation as the breaking-down of the rock encroached on higher land. In other cases the rain-channels have cut in without removing the intervening rocks at once, and formed deep gorges or canyons, which sometimes extend to great distances. They frequently communicate in every direction, forming curious labyrinths, and when the intervening masses are cut away at various levels, or left standing like monuments, we have the characteristic peculiarities of "bad lands" or *mauvaises terres.*

In portions of Kansas tracts of this kind are scattered over the country along the margins of the river and creek valleys and ravines. The upper stratum of the rock is a yellow chalk; the lower, bluish, and the brilliancy of the color increases the picturesque effect. From elevated points the plains appear to be dotted with ruined villages and towns, whose avenues are lined with painted walls of fortifications, churches, and towers, while side alleys pass beneath natural bridges or expand into small pockets and caverns, smoothed by the action of the wind, carrying hard mineral particles.

But this is the least interesting of the peculiarities presented by these rocks. On the level surfaces, denuded of soil, lie huge oyster-shells, some opened and others with both valves together, like remnants of a half-finished meal of some titanic race, who had been frightened from the board, never to return. These shells are not thickened like most of those of past periods, but contained an animal which would have served as a meal for a large party of men. One of them measured twenty-six inches across.

If the explorer searches the bottoms of the rain-washes and ravines, he will doubtless come upon the fragment of a tooth or jaw, and will generally find a line of such pieces leading to an elevated position on the bank or bluff, where lies the skeleton of some monster of the ancient sea. He may find the vertebral column running far into the limestone that locks him in his last prison; or a paddle extended on the slope, as though entreating aid; or a pair of jaws lined with horrid teeth which grin despair on enemies they are helpless to resist. Or he may find a conic mound, on whose apex glisten in the sun the bleached bones of one whose last office has been to preserve from destruction the friendly soil on which he reposed. Sometimes a pile of huge remains will be discovered, which the dissolution of the rock has deposited on the lower level, the force of rain and wash having been insuffìcent to carry them away.

But the reader inquires, What the nature of these creatures thus left stranded a thousand miles from either ocean? How came they in the limestones of Kansas, and were they denizens of land or sea? It may be replied that our knowledge of this chapter of ancient history is only about five years old, and has been brought to light by geological explorations set on foot by Dr. Turner, Prof. Mudge, Prof Marsh, W. E. Webb, and the writer. Careful examinations of the remains discovered show that they are all to be referred to the reptiles and fishes. We find that they lived in the period called Cretaceous, at the time when the chalk of England and the green sand marl of New Jersey were being deposited, and when many other huge reptiles and fishes peopled both sea and land in those quarters of the globe. The twenty-six species of reptiles found in Kansas, up to the present time, varied from ten to eighty feet in length, and represented six orders, the same that occur in the other regions mentioned. Two only of the number were terrestrial in their habits, and three were flyers; the remainder were inhabitants of the salt ocean. When they swam over what are now the plains, the coast-line extended from Arkansas to near Fort Riley, on the Kansas River, and, passing a little eastward, traversed Minnesota to the British Possessions, near the head of Lake Superior. The extent of sea to the westward was vast, and geology has not yet laid down its boundary; it was probably a shore now submerged beneath the waters of the North Pacific Ocean.

Far out on its expanse might have been seen in those ancient days, a huge snake-like form which rose above the surface and stood erect, with tapering throat and arrow-shaped head; or swayed about, describing a circle of twenty feet radius above the water. Then it would dive into the depths, and naught would be visible but the foam caused by the disappearing mass of life. Should several have appeared together, we can easily imagine tall twining forms, rising to the height of the masts of a fishing fleet, or like snakes twisting and knotting themselves together. This extraordinary neck, for such it was, rose from a body of elephantine proportions; and a tail of the serpent pattern balanced it behind. The limbs were probably two pairs of paddles, like those of *Plesiosaurus*, from which this diver chiefly differed in the arrangement of the bones of the breast. In the best known species, twenty-two feet represent the neck, in a total length of fifty feet.

This is the *Elasmosaurus platyurus* (Cope), a carnivorous sea reptile, no doubt adapted for deeper waters than many of the others. Like the snake-bird of Florida, it probably often swam many feet below the surface, raising the head to the distant air for a breath, then withdrawing it and exploring the depths forty feet below, without altering the position of its body. From the localities in which the bones have been found in Kansas, it must have wandered far from land, and that many kinds of fishes formed its food, is shown by the teeth and scales found in the position of its stomach.

A second species, of somewhat similar character and habits, differed very much in some points of structure. The neck was drawn out to a wonderful degree of attenuation, while the tail was relatively very stout, more so, indeed, than in the *Elasmosaurus*, as though to balance the anterior regions while occupied in various actions, *e. g.*, while capturing its food. This was a powerful swimmer, its paddles measuring four feet in length, with an expanse, therefore, of about eleven feet. It is known as *Polycotylus latipinnis* (Cope).

The two species just described formed a small representation, in our great interior sea, of an order which swarmed at the same time, or near it, over the gulfs and bays of old Europe. There they abounded twenty to one. Perhaps one reason for this was the almost entire absence of the real rulers of the waters of Ancient America, viz: the *Pythonomorphs.* These sea-serpents, for such they were, embrace more than half the species found in the limestone rocks in Kansas, and abound in those of New Jersey and Alabama. Only four have been seen as yet in Europe.

Researches into their structure have shown that they were of wonderful elongation of form, especially of tail; that their heads were large, flat, and conic, with eyes directed partly upwards; that they were furnished with two pairs of paddles like the flippers of a whale, but with short or no portion representing the arm. With these flippers and the eel-like strokes of their flattened tail they swam—some with less, others with greater speed. They were furnished, like snakes, with four rows of formidable teeth on the roof of the mouth. Though these were not designed for mastication, and without paws for grasping could have been little used for cutting, as weapons for seizing their prey they were very formidable. And here we have to consider a peculiarity of these creatures in which they are unique among animals. Swallowing their prey entire, like snakes, they were without that wonderful expansibility of throat, due in the latter to an arrangement of levers supporting the lower jaw. Instead of this each half of that jaw was articulated or jointed at a point nearly midway between the ear and the chin. This was of the ball and socket type, and enabled the jaw to make an angle outward, and so widen, by much, the space inclosed between it an its fellow. The arrangement may be easily imitated by directing the arms forward, with the elbows turned outward and the hands placed near together. The ends of these bones were in the Pythonomorphs as independent as in the serpents, being only bound by flexible ligaments. By turning the elbows outward, and bending them, the space between the arms becomes diamond-shaped, and represents exactly the expansion seen in these reptiles, to permit the passage of a large fish or other body. The arms, too, will represent the size of jaws attained by some of the smaller species. The outward movement of the basal half of the jaw necessarily twists in the same direction the column-like bone to which it is suspended. The peculiar shape of the joint by which the last bone is attached to the skull, depends on the degree of twist to be permitted, and, therefore, to the degree of expansion of which the jaws were capable. As this differs much in the different species, they are readily distinguished by the column or "quadrate" bone when found. There are some curious consequences of this structure, and they are here explained as an instance of the mode of reconstruction of extinct animals from slight materials. The habit of swallowing large bodies between the branches of the under-jaw necessitates the prolongation forward of the mouth of the gullet; hence the throat in the Pythonomorphs must have been loose and almost as baggy as a pelican's. Next, the same habit must have compelled the forward position of the glottis or opening of the windpipe, which is always in front of the gullet. Hence these creatures must have uttered no other sound than a hiss, as do animals of the present day which have a similar structure, as for instance, the snakes. Thirdly, the tongue must have been long and forked and for this reason: its position was still anterior to the glottis, so that there was no space for it except it were inclosed in a sheath beneath the windpipe when at rest, or thrown out beyond the jaws when in motion. Such is the arrangement in the nearest living forms, and it is always, in these cases, cylindric and forked.

The flying saurians of the cretaceous sea of Kansas, though not numerous in species, were of remarkable size. Though their remains are generally flattened by the pressure of the overlying rocks, two species have left a complete record of their form and dimensions. One of them (*Ornithochirus Tarpyia*) spread eighteen feet between the tips of the wings, while the *O. umbrosus* covered nearly twenty-five feet with his expanse. These strange creatures flapped their leathery wings over the waves, and, often plunging, drew many a fish from its companions of the shoal; or, soaring at a safe distance, viewed the sports and combats of the more powerful saurians of the sea; or, trooping to the shore at nightfall, suspended themselves to the cliffs by the claw-bearing fingers of their wing-limbs.

DEVELOPING—ONE OF THE FIRST FAMILIES.

In connection with the subject of the old lakes and their fertile shores, where human beings, it might reasonably be expected, once lived so comfortably, the editor of this volume begs to lay before the reader (in a sort of parenthesis, for which Professor Cope is in no way responsible) an effort of Sachem's. He dedicated it to Darwin, and was pleased to call it, notwithstanding it smells more of the fossil-bone caves than the fields,

THE PRIMEVAL MAN'S PASTORAL.

My grandfather Jock was an ape,
His grandfather Twist was a worm;
Each age has developed in shape,
And ours has got rid of the squirm;
If the law of selection will work in our case,
We 'll develop, in time, to a wonderful race.

My sweetheart has claws, and her face
Is covered with bristles and hair;
She 's feline in nature and grace,
She 's apt to get out on a tear,
She 's cursed with a passion to sing after night;
But these she 'll evolve, and develop all right.

One race has evolved in the sea,
And partly got rid of their scales;
Though cousin by faces to me,
They 're cousin to fishes by tails;
But they 'll ever remain simply mer-men and women,
For selection won't work, in the world that they swim in.

'T is said that Gorilla the Great,
  Who rules as the chief of our clan,
Has found in the annals of fate,
  We 're soon to evolve into man ;
Furthermore, that our children will doubt whence they came,
  Till a fellow named Darwin shall put them to shame.

CHAPTER XXIV.

CONTINUED BY COPE—THE GIANTS OF THE SEAS—TAKING OUT FOSSILS IN A GALE—INTERESTING DISCOVERIES—THE GEOLOGY OF THE PLAINS.

THE giants of the Pythonomorphs of Kansas have been called *Liodon proriger* (Cope) and *Liodon dyspelor* (Cope). The first must have been abundant, and its length could not have been far from sixty feet, certainly not less. Its physiognomy was rendered peculiar by a long projecting muzzle, reminding one of that of the blunt-nosed sturgeon of our coast, but the resemblance was destroyed by the correspondingly massive end of the branches of the lower jaw. Though clumsy in appearance, such an arrangement must have been effective as a ram, and dangerous to his enemies in case of collision. The writer once found the wreck of an individual of this species strewn around a sunny knoll beside a bluff, and his conic snout, pointing to the heavens, formed a fitting monument, as at once his favorite weapon, and the mark distinguishing all his race.

Very different was the *Liodon dyspelor*, a still larger animal than the last, with a formidable armature. It was indeed the longest of known reptiles, and probably equal to the great finner whale of modern oceans. The circumstances attending the discovery of one of these, will always be a pleasant recollection to the writer. A part of the face, with teeth, was observed projecting from the side of a bluff by a companion in exploration, (Lieut. Jas. H. Whitten, U. S. A.), and we at once proceeded to follow up the indication with knives and picks. Soon the lower jaws were uncovered, with their glistening teeth, and then the vertebræ and ribs. Our delight was at its height when the bones of the pelvis and part of the hind limb were laid bare, for they had never been seen before in the species and scarcely in the order. While lying on the bottom of the cretaceous sea, the carcass had been dragged hither and thither by the sharks and other rapacious animals, and the parts of the skeleton were displaced and gathered into a small area. The massive tail stretched away into the bluff, and after much laborious excavation we left a portion of it to more persevering explorers. The species of *Clidastes* did not reach such a size as some of the *Liodons*, and were of elegant and flexible build. To prevent their habits of coiling from dislocating the vertebral column, these had an additional pair of articulations at each end, while their muscular strength is attested by the elegant striæ and other sculptures which appear on all their bones. Three species of this genus occur in the Kansas strata, the largest (*Clidastes cineriarum*, Cope) reaching forty feet in length. The discovery of a related species (*Holcodus coryphæus*, Cope) was made by the writer under circumstances of difficulty peculiar to the plains. After examining the bluffs for half a day without result, a few bone fragments were found in a wash above their base. Others led the way to a ledge forty or fifty feet from both summit and foot, where, stretched along in the yellow chalk, lay the projecting portions of the whole monster. A considerable number of vertebræ were found preserved by the protective embrace of the roots of a small bush, and, when they were secured, the pick and knife were brought into requisition to remove the remainder. About this time one of the gales, so common in that region, sprang up, and, striking the bluff fairly, reflected itself upwards. So soon as the pick pulverized the rock, the limestone dust was carried into eyes, nose, and every available opening in the clothing. I was speedily blinded, and my aid disappeared into the canyon, and was seen no more while the work lasted. Only the enthusiasm of the student could have endured the discomfort, but to him it appeared a most unnecessary "conversion of force" that a geologist should be driven from the field by his own dust. A handkerchief tied over the face, and pierced by minute holes opposite the eyes, kept me from total blindness, though dirt in abundance penetrated the mask. But a fine relic of creative genius was extricated from its ancient bed, and one that leads its genus in size and explains its structure.

On another occasion, riding along a spur of a yellow chalk bluff, some vertebræ lying at its foot met my eye. An examination showed that the series entered the rock, and, on passing round to the opposite side the jaws and muzzle were seen projecting from it, as though laid bare for the convenience of the geologist. The spur was small and of soft material, and we speedily removed it in blocks, to the level of the reptile, and took out the remains, as they laid across the base from side to side. A genus related to the last is *Edestosaurus*. A species of thirty feet in length, and of elegant proportions has been called *E. tortor* (Cope.) Its slenderness of body was remarkable, and the large head was long and lance-shaped. Its flippers tapered elegantly, and the whole animal was more of a serpent than any other of its tribe. Its lithe movements brought many a fish to its knife-shaped teeth, which are more efficient and numerous than in any of its relatives. It was found coiled up beneath a ledge of rock, with its skull lying undisturbed in the center. A species distinguished for its small size and elegance is *Clidastes pumilus* (Marsh). This little fellow was only twelve feet in length, and was probably unable to avoid occasionally furnishing a meal for some of the rapacious fishes which abounded in the same ocean.

Tortoises were the boatmen of the cretaceous waters of the eastern coast, but none had been known from the deposits of Kansas until very recently. One species now on record (*Protostega gigas*, Cope), is of large size, and strange enough to excite the attention of naturalists. It is well known that the house or boat of the tortoise or turtle is formed by the expansion of the usual bones of the skeleton till they meet and unite, and thus become continuous. Thus the lower shell is formed of united ribs of the breast and breast-bone, with bone deposited in the skin. In the same way the roof is formed by the union of the ribs with bone deposited in the skin. In the very young tortoise the ribs are separate as in other animals; as they grow older they begin to expand at the upper side of the upper end, and, with increased age, the expansion extends throughout the length. The ribs first come in contact where the process commences, and in the land-tortoise they are united to the end. In the sea-turtle, the union ceases a little above the ends. The fragments of the *Protostega* were seen by one of the men projecting from a ledge of a low bluff. Their thinness and the distance to which they were traced excited my curiosity, and I straightway attacked the bank with the pick. After several square feet of rock had been removed, we cleared up one floor, and found ourselves well repaid. Many long slender pieces, of two inches in width, lay upon the ledge. They were evidently ribs, with the usual heads, but behind each head was a plate like the flattened bowl of a huge spoon placed crosswise. Beneath these stretched two broad plates two feet in width, and no thicker than binders' board. The edges were fingered, and the surface hard and smooth. All this was quite new among fully grown animals, and we at once determined that more ground must be explored, for further light. After picking away the bank and carving the soft rock, new masses of strange bones were disclosed. Somę bones of a large paddle were recognized, and a leg bone. The shoulder-blade of a huge tortoise came next, and further examination showed that we had stumbled on the burial-place of one of the largest species of sea-turtle yet known. The single bones of the paddle were eight inches long, giving the spread of the expanded flippers as considerably over twenty feet. But the ribs were those of an ordinary turtle just born, and the great plates represented the bony deposit in the skin, which, commencing independently in modern turtles, united with the expanded ribs below, at an early day. But it was incredible that the largest of known turtles should be but just hatched, and for this and other reasons it has been concluded that this "ancient mariner" is one of those forms not uncommon in old days, whose incompleteness in some respects points to the truth of the belief, that animals have assumed their modern perfection, by a process of growth from more simple beginnings.

The cretaceous ocean of the West was no less remarkable for its fishes than for its reptiles. Sharks do not seem to have been so common as in the old Atlantic, but it swarmed with large predaceous forms related to the salmon and saury.

Vertebræ and other fragments of these species project from the worn limestone in many places. I will call attention to, perhaps, the most formidable, as well as the most abundant of these. It is the one whose bones most frequently crowned knobs of shale, which had been left standing amid surrounding destruction. The density and hardness of the bones shed the rain off on either side, so that the radiating gutters and ravines finally isolated the rock mass from that surrounding. The head was some inches longer than that of a fully grown grizzly bear, and the jaws were deeper in proportion to their length.

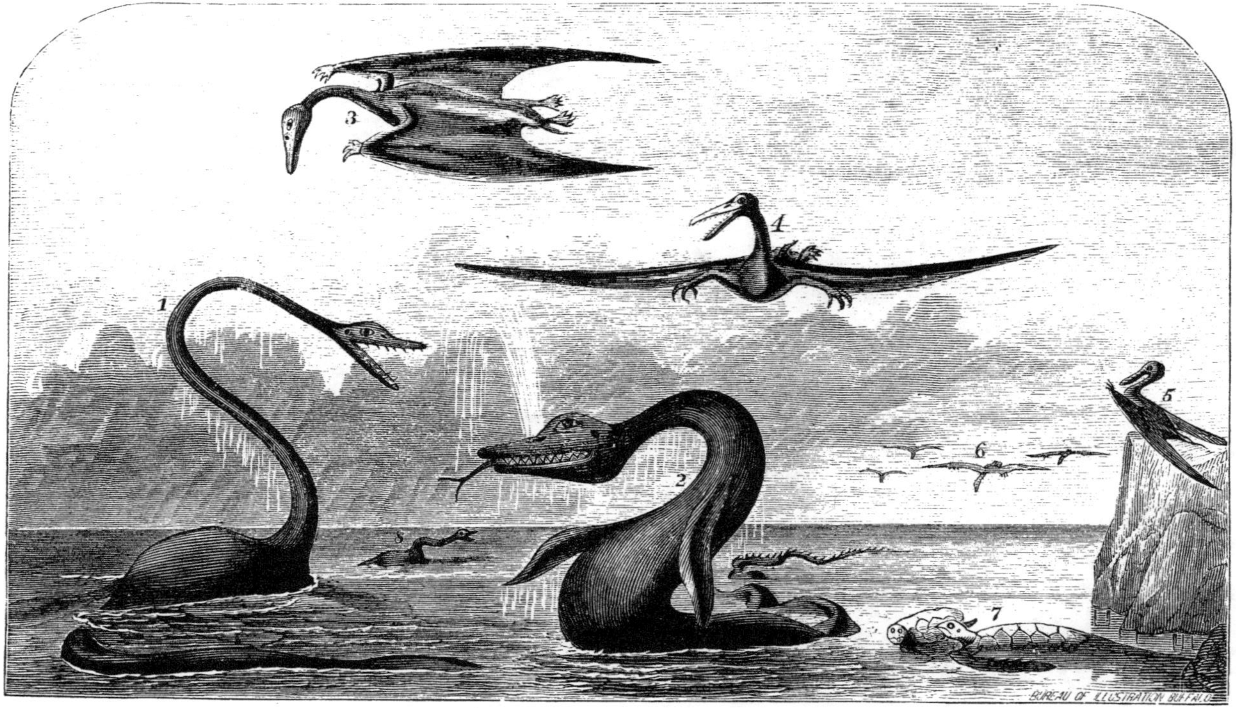
1
2
3
4
5
6
7
8
BUREAU OF ILLUSTRATION BUFFALO

THE SEA WHICH ONCE COVERED THE PLAINS.

Elasmosaurus platyurus. 2. Liodon proriger. 3, 4, 5. Ornithochirus umbrosus. 6. Ornithochirus harpyia. 7. Protostega jigas. 8. Polycotylus latipinnis.

The muzzle was shorter and deeper than that of a bull-dog. The teeth were all sharp cylindric fangs, smooth and glistening, and of irregular size. At certain distances in each jaw they projected three inches above the gum, and were sunk one inch into the bony support, being thus as long as the fangs of a tiger, but more slender. Two such fangs crossed each other on each side of the middle of the front. This fish is known as *Portheus molossus* (Cope). Besides the smaller fishes, the reptiles no doubt supplied the demands of his appetite.

The ocean in which flourished this abundant and vigorous life, was at last completely inclosed on the west, by elevation of sea-bottom, so that it only communicated with the Atlantic and Pacific at the Gulf of Mexico and the Arctic Sea. The continued elevation of both eastern and western shores contracted its area, and when ridges of the sea-bottom reached the surface, forming long low bars, parts of the water area were inclosed and connection with salt water prevented. Thus were the living beings imprisoned and subjected to many new risks to life. The stronger could more readily capture the weaker, while the fishes would gradually perish through the constant freshening of the water. With the death of any considerable class the balance of food supply would be lost, and many larger species would disappear from the scene. The most omnivorous and enduring would longest resist the approach of starvation, but would finally yield to inexorable fate; the last one caught by the rising bottom among shallow pools from which his exhausted energies could not extricate him.

PART II—GEOLOGY.

The geology of this region has been very partially explored, but appears to be quite simple. The following description of the section along the line of the Kansas Pacific Railroad, will probably apply to similar sections north and south of it. The formations referable to the cretacious period on this line, are those called by Messrs. Meek and Hayden the Dakota, Benton, and Niobrara groups, as Nos. 1, 2 and 3. According to Leconte,* at Salina, one hundred and eighty-five miles west of the State line of Missouri, the rocks of the Dakota group constitute the bluffs, and continue to do so as far as Fort Harker, thirty-three miles farther west. They are a "coarse brown sand-stone, containing irregular concretions of oxide of iron," and numerous molluscs of marine origin. Near Fort Harker, certain strata contain large quantities of the remains (leaves chiefly) of dicotyledonous and other forms of land vegetation. Near this point, according to the same authority, the sand-stone beds are covered with clay and limestone. These he does not identify, but portions of it from Bunker Hill, thirty-four miles west, have been identified by Dr. Hayden, as belonging to the Benton or second group. The specimen consisted of a block of dark, bluish-gray clay rock, which bore the

* Notes on the geology of the survey for the extension of the Union Pacific Road E. D. from the Smoky Hill to the Rio Grande, by John L. Leconte, M. D. Philadelphia, 1868.

remains of the fish *Apsopelix sauriformis* (Cope). That the eastern boundary of this bed is very sinuous is rendered probable by its occurrence at Brookville, eighteen miles to the eastward of Fort Harker, on the railroad. In sinking a well at this point, the same soft, bluish clay rock was traversed, and at a depth of about thirty feet a skeleton of a saurian of the crocodilian order was encountered, the *Hyposaurus vebbii* (Cope).

The boundary line, or first appearance of the beds of the Niobrara division, has not been pointed out, but at Fort Hays, seventy miles west of Fort Harker, its rocks form the bluffs and outcrops every-where. From Fort Hays to Fort Wallace, near the western boundary of the state, one hundred and thirty-four miles beyond, the strata present a tolerably uniform appearance. They consist of two portions; a lower, of dark-bluish calcareo-argillaceous character, often thin-bedded; and a superior, of yellow and whitish chalk, much more heavily bedded. Near Fort Hays the best section may be seen, at a point eighteen miles north, on the Saline river. Here the bluffs rise to a height of two hundred feet, the yellow strata constituting the upper half. No fossils were observed in the blue bed, but some moderate-sized *Ostreæ*, frequently broken, were not rare in the yellow. Half way between this point and the Fort, my friend, N. Daniels, of Hays, guided me to a denuded tract, covered with the remains of huge oysters, some of which measured twenty-seven inches in diameter. They exhibited concentric obtuse ridges on the interior side, and a large basin-shaped area behind the hinge.

Fragments of fish vertebræ of *Anogmius* type were also found here by Dr. Janeway. These were exposed in the yellow bed. Several miles east of the post, Dr. J. H. Janeway, Post Surgeon, pointed out to me an immense accumulation of *Inoceramus problematicus* in the blue stratum. This species also occurred in abundance in the bluffs west of the Fort, which were composed of the blue bed, capped by a thinner layer of the yellow. Large globular or compound globular argillaceous concretions, coated with gypsum, were abundant at this point.

Along the Smoky Hill River, thirty miles east of Fort Wallace, the south bank descends gradually, while the north bank is bluffy. This, with other indications, points to a gentle dip of the strata to the north-west. The yellow bed is thin or wanting on the north bank of the Smoky, and is not observable on the north fork of that river for twenty miles northward or to beyond Sheridan Station, on the Kansas Pacific Railroad. Two isolated hills, "The Twin Buttes," at the latter point are composed of the blue bed, here very shaly to their summits. This is the general character of the rock along and north of the railroad between this point and Fort Wallace.

South of the river the yellow strata are more distinctly developed. Butte Creek Valley, fifteen to eighteen miles to the south, is margined by bluffs of from twenty to one hundred and fifty feet in height on its southern side, while the northern rises gradually into the prairie. These bluffs are of yellow chalk, except from ten to forty feet of blue rock at the base, although many of the canyons are excavated in the yellow rock exclusively. The bluffs of the upper portion of Butte Creek, Fox, and Fossil Spring (five miles south) canyons, are of yellow chalk, and report of several persons stated that those of Beaver Creek, eight miles south of Fossil Spring, are exclusively of this material. Those near the mouth of Beaver Creek, on the Smoky, are of considerable height, and appear at a distance to be of the same yellow chalk.

I found these two strata to be about equally fossilliferous, and I am unable to establish any palæontological difference between them. They pass into each other by gradations in some places, and occasionally present slight laminar alternations at their line of junction. I have specimens of *Cimolichthys semianceps* (Cope), from both the blue and yellow beds, and vertebrae of the *Liodon glandiferus* (Cope) were found in both. The large fossil of *Liodon dyspelor* (Cope) was found at the junction of the bed, and the caudal portion was excavated from the blue stratum exclusively. Portions of it were brought East in blocks of this material, and these have become yellow and yellowish on many of the exposed surfaces. The matrix adherent to all the bones has become yellow. A second incomplete specimen, undistinguishable from this species, was taken from the yellow bed.

As to mineral contents, the yellow stratum is remarkably uniform in its character. The blue shale, on the contrary, frequently contains numerous concretions, and great abundance of thin layers of gypsum and crystals of the same. Near Sheridan concretions and septaria are abundant. In some places the latter are of great size and, being embedded in the stratum, have suffered denudation of their contents, and, the septa standing out, form a huge honey-comb. This region and the neighborhood of Eagle Tail, Colorado, are noted for the beauty of their gypsum-crystals, the first abundantly found in the cretaceous formation. These are hexagonal-radiate, each division being a pinnate or feather-shaped lamina of twin rows of crystals. The clearness of the mineral, and the regular leaf and feather forms of the crystals give them much beauty. The bones of vertebrate fossils preserved in this bed are often much injured by the gypsum formation which covers their surface and often penetrates them in every direction.

The yellow bed of the Niobrara group disappears to the south-west, west, and north-west of Fort Wallace, beneath a sandy conglomerate of uncertain age. Its color is light, sometimes white, and the component pebbles are small and mostly of white quartz. The rock wears irregularly into holes and fissures, and the soil covering it generally thin and poor. It is readily detached in large masses, which roll down the bluffs. No traces of life were observed in it, but it is probably the eastern margin of the southern extension of the White River Miocene Tertiary stratum. This is at least indicated by Dr. Hayden, in his geological preface to Leidy's extinct mammals of Dakota and Nebraska.

Commercially, the beds of the Niobrara formation possess little value, except when burned for manure.

The yellow chalk is too soft in many places for buildings of large size, but will answer well for those of moderate size. It is rather harder at Fort Hays, as I had occasion to observe at their quarry. That quarried at Fort Wallace does not appear to harden by exposure; the walls of the hospital, noted by Leconte on his visit, remained in 1871 as soft as they were in 1867. A few worthless beds of bituminous shale were observed in Eastern Colorado.

The only traces of Glacial Action in the line explored were seen near Topeka. South of the town are several large, erratic masses of pink and bloody quartz, whose surfaces are so polished as to appear as though vitrified. They were transported, perhaps, from the Azoic area near Lake Superior.

CHAPTER XXV.

A SAVAGE OUTBREAK—THE BATTLE OF THE FORTY SCOUTS—THE SURPRISE—PACK-MULES STAMPEDED—DEATH ON THE ARICKEREE—THE MEDICINE MAN—A DISMAL NIGHT—MESSENGERS SENT TO WALLACE—MORNING ATTACK—WHOSE FUNERAL?—RELIEF AT LAST—THE OLD SCOUTS' DEVOTION TO THE BLUE.

ON our return to Sheridan we were deeply pained to hear of the sad death of Doctor Moore and Lieutenant Beecher, whose acquaintance we had formed at Fort Hays, and the former of whom we had learned to esteem most highly as a personal friend. A scouting party, not long before, had left the post just named, under the command of General Forsythe, of Sheridan's staff, and composed principally of those citizens who had seen frontier service. Dr. Moore accompanied it as surgeon, and Lieut. Beecher—a nephew of Henry Ward Beecher, and an officer of the regular army—held the position of chief of scouts, which he had filled for some time previously with much credit. The savages of the plains being again upon the war-path, that brave and well-organized little party of fifty were dispatched to pursue a band of Indians, which had appeared before Sheridan and run off a lot of stock.

Some of the scouts were now in the town, and from

one of them we obtained an account of the expedition. Fresh from the mouth of that sandy hell in the river's head, which had sucked out the life-blood of so many of his companions, I wish my readers could have heard the story told with the rude eloquence in which he clothed it. As it is, how nearly they will come to doing so, must perforce depend on how nearly I can remember his language.

"You see, captain," he began (it is considered impolite among this class ever to address one without using some title), "we had the nicest little forty lot o' scouts that ever followed the plains fur a living, and trails fur an Injun. Thar wur ingineers, doctors, counter-jumpers, and a few deadbeats, but every one of 'em had lots of fight, and not the least bit of scare. Ther talents run ter fightin', an' ther bodies never run away from it.

"It wur kinder curious, though, to see the chaps that wur not bred ter ther business git along. They wur the profession folks. Some had a little compass, not much bigger 'n a button, that they carried on the sly. Good scouts do n't need no such fixin's. These uns 'ud reach inter ther pockets, as if they was going ter take a chaw o' terbaccer, and gettin' a sly wink at ther needle, would cry out ter ther neighbors, 'I say, hoss, we 're goin' a little too much east of north!' or, 'I tell yer what, fel, we 're at least two p'ints off our course.' And all ther time they could n't have told south from west, without them needles. But ther war n't a coward in the whole pack. Every one had a back as stiff fur a fight as a cat.

"We struck a large Injun trail the fourth day out,

and kept it till evenin', but no other sign showed it self over ther wide reach that would have told a livin' bein' had ever bin thar before us. Next mornin', early, ther was a sudden fuss among our horses, and a cry from the guard, and, afore we knew it, eight pack-mules had been stampeded, and driven off. It wur a narrow call fur ther whole herd.

"The fellers had come down a ravine until they got close enough, and, then suddenly rushin' along in the grayness, set the mules inter a crazy run, and gathered 'em up, out of gun-shot. You may lick a pack-mule along all day, and be afraid he 'll drop down dead, and yet give him a fair chance to stampede, and he 'll outrun an elk, and grow fat on it.

"Stock and Injuns was both out of sight in a jiffy, and the order was given to saddle, and recapture. We were just raisin' inter ther stirrups, when some of the boys called out, and we saw the whole valley ahead of us filled with Injuns comin' down. Ther war n't no mules lost just then, and we kinder fell back onto a sort of high-water island in the Arickeree. That, yer know, is the dry fork of the Republican. Bein' low water then, as it is most of the time thar, nothin' but a dry bed of sand was on each side.

"It seemed as if the whole Injun nation was coming down on us. Such a crowd o' lank ponies, and painted heathen astride, yer never see. I expected seein' of 'em would prevent *my* ever seein' of my family agin. 'Jim,' says I to my chum, and 'Bill,' says he to me, and then we did n't say nothin' more, but as the heathen come a chargin', we both put a hand in our pockets, just as if the brains had been in one head, and then both of us took a chaw o' ter-baccer.

"For the next few hours ther wur an awful scrim-mage, and a shootin', and a hollerin', and a whizzin' of bullets, which made that the hottest little island ever stranded on sand. The boys had all dug out, with their hands, sort o' little rifle-pits, and fit behind 'em. We had good Spencers, with a few Henrys, and the way those patents spit lead at the devils' hearts wur a caution. The first charge, they cum close up to us, and for a hull minnit, that stretched out awfully, we were afraid they 'd ride us down. It was reg'lar coffee-mill work then, grindin' away at the levers, and we flung bullets among 'em astonish-in'. As fast as one Injun keeled, another 'd pick him up, and nary dead was left on the field.

"They follered up the charge game by a siege one, and peppered away at us from the neighborin' ravines and hills. Ther number wur about eight hundred, and some had carbines, and others old rifles and pistols. A few would sneak along in the bottom grass, and get behind trees, and then thur would be a flash, and a crack, and the ball would come tearin' in among us, sometimes burrowin' in a human skull, or elsewise knockin' down a horse. And all around, on the ridges, the squaws were a dancin' and shoutin', and the braves, whenever any of 'em got tired of shootin', would join their ugly she's, and help 'em in kickin' up a hullabaloo.

"I reckon, arter they 'd killed the last hoss, they must ha' had a separate scalp-dance fur each one on us. Plain sailin' then, ther red fellows thought—less than fifty white men down in the sand, and most a thousan' Injuns roun' 'em, and more 'n a hundred miles to the nearest fort; the weaker party bein' afoot, too, and the other mounted.

"But we soon made 'em pitch another tune, beside ther juberlatin' one. We had took notice of a big Injun, with lots o' fixins on him, cavortin' all round ther island, and a spurrin' up the braves. We made certain it wur the medicine man, and found out arterward that he 'd been tellin' on 'em ther pale-faces' bullets would melt before reachin' an Injun. Six on us got our rifles together, and as ther old copper-colored Pillgarlic cum dancin' round, we let fly. If Injun carcasses go along with ther spirits, I reckon ther bullets we put into the old sinner, got melted, sure enough. And what a howlin' thur was, as his pony scampered in among the squaws, empty saddled!

"It wur an awful sight to look roun' among our little sand-works—twenty killed and wounded men, covered with blood and grit. Our leader, Col. Forsythe, was shot in both legs, a ball passin' through the thigh part of one, and a second breakin' the bones of the other below the knee. He wur a knowin' and cool officer.

"Lieut. Beecher, a nephew of the big preacher, was shot through the small o' the back, and lay thar beggin' us to kill him. He too wur a brave man, and did n't flinch, never, from duty nor danger. They say that his two sisters were drowned from a sail-boat on the Hudson, two years ago, and that the old parents are left now all alone. Doc. Moore was shot through the head, and sat thar noddin', and not knowin' no one. I spoke to him once, and he kinder started back, as if he see the Injun which shot him, still thar. He wur a good surgeon, and all the boys liked him. I hev got his gun down at my tent, all full o' sand, whar it got tramped arter he fell. *

"Culver lay dead on one side of our little island, shot by an Injun that crawled up in the grass. Lots o' others was wounded, and our chances looked as dark as ther night which wur coming down on us. But we was glad ter see daylight burn out, as it kinder gin us a chance to rest and think.

"That night was awful dismal. The little spot o' sand, down thar in the river's bed, seemed ther only piece o' earth friendly to us, and we were clingin' to it like sailors ter a raft at sea. The darkness all around was a gapin' ter swaller us, and a hidin' its blood-hounds, to set 'em on with ther sun. Night, without any thin' in it more 'n grave-stones, is terrifyin' to most people, but just you fill it full of pantin's for blood in front, and Death sittin' behind, among the corpses, and watchin' the wounded, and a feller's blood falls right down to January. It kinder thickens, like water freezin' round the edges, and your hands and feet get powerful cold, and you feel as if you would n't ever be thawed out, this side of the very place you don't want ter go to.

"Toward midnight, Stillwell and Trudell crawled out o' camp, to go for relief. They were to creep and sneak through the Injun lines, and get beyond

* I obtained the weapon that I had loaned our friend, and have carefully kept it since, as a memento.

'em by daylight. Then they would lay by, and push on ag'in, when dark cum, toward Wallace. That little spot of barracks, a hundred and twenty-five miles off, kept up our hope mightily. It was our lighthouse, like. We were shipwrecked among savages, and had sent a couple of yawls off, to tell the keeper thar of danger. We knew if the news reached, blue coats would flash out to us, like spots of light, and our foes go before 'em as mist.

"But footin' it nights, and layin' by days, fur over a hundred miles, through Injun country, is slow work, and we did n't, most on us, expect much; and our hearts followed the little black spots, showin' us our two companions a creepin' off into darkness, like a couple of wolves. It took good men, too, from our little party, and fur awhile I was faint-hearted. In our shipwreck, it seemed like takin' bottles which might ha' helped to hold out, and flingin' 'em into ther waves, with messages tellin' how and whar we went down.

"About two o'clock Lieut. Beecher died, havin' for some time begged the men to end his sufferin's by shootin' of him.

"We all kept perfect quiet that night—no fire, nor wur ther a sound heard, from our little island, by the heathen on the bluffs. An just that quietness gave 'em the worst foolin' they ever had. It seems the road down river had been left open by 'em, hopin' we would steal out and run for it durin' the night. We bein' all on foot, they could overtake us in the mornin', and worry on us out easy. Durin' the dark we waited quiet, and watched, and passed water to our wounded, and sprinkled it over some of 'em who could n't drink.

"It wer just kinder palin' like way up in the sky, and we could see that off down East, somewhar, ther mornin' was commencin' ter climb, when Jim nudged me, and says, 'Chum, what 's that?' We both stuck our ears right up, like two jackass-rabbits, and listened. It wur all dark near the ground, but we could hear a steady, gallopin' sound, comin' in toward us from up the ravines, and over the hills. It wur like a beatin' of ther earth with flails by threshers you could n't see.

"The sound came a creepin' along the sod so quick we soon knew it wur the Injuns, on ther ponies, comin' down ter pick up the trail. And now we could see 'em a bobbin' along toward us in ther gloom, the rows er ugly heads goin' up and down, like jumpin'-jacks. It just seemed as ther side er ther night had been painted all full o' gapin' red devils, and ther sun wur jest revealin' on 'em. 'Lay still!' wer the word, and each man hugged his sand bank, just a skinnin' one eye, like a lizard over a log. They 'd no idee we were thar, not bein' able to understand the grit of that little forty, and they cum gallopin' along, careless-like, happy as so many ghosts goin' ter a fun'ral. But it war n't *our* fun'ral just then. When they 'd got so close we could smell 'em, colonel guv the word ter fire, and we let 'em have it. Stranger, you ain't no idee what a gettin' up bluffs, and general absentin' of 'emselves ther wur. Arter the fust crack, yer could n't see an Injun at all, but jest a lot er ponies, diggin' it on ther back track, and you knowed painted cusses wer glued ter ther opposite side on 'em.

"We had fightin' until night ag'in, but no men were killed arter the fust day. The savages were cautious-like, and took long range fur it. We now commenced cuttin' off the hind quarters of our dead hosses, and boilin' small pieces in a empty pickle-jar belongin' ter ther colonel. Burke, he 'd dug a shallow well, too, which gave us plenty of water. Hoss meat is n't relishin' at fust. One kin eat it, but, as ther feller said about crow, he don't hanker arter it. Ther gases had got all through ther carcasses, and we had ter sprinkle lots o' gunpowder inter the pot, to kill the taste.

"The fust hoss cut up was my old sorrel. He did n't go well while livin', and could n't be expected to when dead. Instead of takin' a straight course, and givin' some satisfaction, he jumped across all the turns inside o' me, and brought up bump agin my hide, as if he wer comin' through. He had that same trick o' cuttin' corners when livin', and I perceded ter give him up as a uncontrollable piece of hoss flesh.

"When night come on agin, Pliley and Whitney attempted ter get through ther Injun lines and make fur Wallace, but were driven back. Fur ther next few days we kept eatin' hoss flesh, and fightin' occasionally. The third night Pliley and Donovan succeeded in gettin' away.

"On the fourth day, Doctor Moore died. After the fifth, no Injuns was visible, and we gathered prickly pears and eat 'em, boilin' some down inter syrup.

Our mouths were all full of ther little needles, and it wer mighty hard keepin' a stiff upper lip. We were eatin' away on our forty-eight horses, and watchin' and hopin'. We could n't move, and leave our wounded, or the Injuns would be on 'em right off. The poor fellows had no surgeon, and were sufferin' terrible as 't was.

"Ther mornin' of ther ninth day broke with a cry of 'Injuns!' Now, human natur' can't stand fitin' allers. To carry out my shipwreck idee, fellers on a raft kin cling an' swaller water fur awhile, but they can't fight a hull grist o' hurricanes. Hoss meat an' prickly pears ain't jest ther thing, either, to slap grit inter a man. Ther were a big crowd comin', sure enough, way off on ther hills. We were kinder beginnin' ter despond, when a familiar sort o' motion on the fur dark line spelt in air the word, 'Friend!' It wer the advanced guard o' relief, approachin' on ther jump. Why, boy"—and the old scout seized hold of Semi, and shook him in excitement—"talk of Lucknow and ther camels a comin', they war n't nowhar. The blessed old blue cloth! If yer want ter love a color, jest get saved by it once. When I get holed in ther earth, I 'll take back ter dust on a blue blanket, an' if I get married afore, gal an' I 'll wear blue, an' the preacher 'll hev ter swar a blue streak in jinin' us!"

We afterward met others of the scouts—intelligent, clear-headed fellows, with much more of cultivation than our rough friend possessed—and they corroborated his story in every particular. I have let him tell it in his own way, not only because vastly more graphic than any words of mine could be, but also to the end that the reader might become acquainted with a genuine frontiersman—one of that class which is wheeling into line with the immense multitudes of Indians and buffalo that time and civilization are bearing swiftly onward to hide among the memories of the past.

That the savages suffered very severely in their several attacks upon that little band of heroes on the Arickeree, was evident from the number of bodies found by the relief, as it hastened forward from Fort Wallace. The corpses were resting on hastily-constructed scaffolds, and some had evidently been placed there while dying, as the ground underneath was yet wet with blood.

CHAPTER XXVI.

THE STAGE DRIVERS OF THE PLAINS—OLD BOB—"JAMAICA AND GINGER"—AN OLD ACQUAINTANCE—BEADS OF THE PAST—ROBBING THE DEAD—A LEAF FROM THE LOST HISTORY OF THE MOUND BUILDERS—INDIAN TRADITIONS—SPECULATIONS—ADOBE HOUSES IN A RAIN—CHEAP LIVING—WATCH TOWERS.

THE stage drivers of the plains are rapidly becoming another inheritance of the past, pushed out of existence by the locomotive, whose cow-catcher is continually tossing them from their high seats into the arms of History. What a rare set they are, though! No two that I ever saw were nearly alike, and they resemble not one distinctive class, but a number. The Jehus who crack their whips over the buffalo grass region, and turn their leaders artistically around sharp corners in rude towns, are made up on a variety of patterns. Some are loquacious and others silent, and while a portion are given to profanity, another though smaller number are men of very proper grammar. Some with whom I have ridden would discount truth for the mere love of the exercise, while others I have found so particular that they could not be induced to lie, except when it was for their interest to do so.

In a village on the shores of Lake Champlain, in the frozen regions of northern New York, where mercury becomes solid in November, and remains so until May, I got on intimate terms, when a boy, with a stage driver. During the long winters the coaches were placed on sleds, and well do I remember the style in which "Old Bob," as he was universally called, would come dashing into the town on frosty mornings, winding uncertain tunes out of a brass horn, given him years before by a General Somebody, of the State Militia. In front of the long-porched tavern, the leaders would push out to the left, in order to give due magnificence to the right hand circle, which deposited the coach at the bar-room door. Bearish in fur, and sour in face, Bob would then roll from the seat, rush up to the bar, and for the first time open his mouth, to ejaculate, "Jamaica and ginger!" The fiery draught would thaw out his tongue, as hot water does a pump, and after that it was easy work to pump him dry of any and all news on the line above.

That was many years ago, and in a spot half a continent away. One morning, while at Sheridan, I heard the blast of a horn up the street, whose notes awakened echoes which had long lain dead and buried in boyhood's memory. A moment more, and out from an avenue of saloons the overland stage rattled, and on its box sat the friend of my childhood, "Old Bob." He had the identical horn, and it was the identical tune, which I had so often heard in the by-gone years, the only difference being that both were cracked, and the lungs behind the mouth-piece, touched by the winters of sixty-odd, wheezed a little. As the coach came to the door, I jumped up by the "boot," and grasping the old fellow's hand, introduced myself. Old Bob rubbed his eyes, which were weak and watery, and scanned me closely.

"Well, well, lad," he said, "your face takes me now, sure enough. I mind your father and mother well, and you're the little rascal that stole my whip once, when I was thawing out with Jamaica and ginger. Did you tell me by the old tune? You did, eh? Well, truth is, lad, the horn won't blow any other. It's got to running in that groove, and when I try to coax any thing new out, it sets off so that it frightens the horses."

The coach was now ready for starting, and, as he gathered up the reins, my friend of auld lang syne called out to me, "When you get back to York State, if you see any Rouse's Point people that ask for Old Bob, tell them he does n't take any Jamaica and ginger now. Tell them he's out on the plains, tryin' to get back some of the life the cussed stuff burnt out of him." And away the stage coach rattled, and soon was out of hearing.

Next day's down stage brought intelligence that Bob's coach had been attacked by Indians, but the old fellow had handled his lines right skillfully, and brought mails and passengers through in safety.

Our last day at Sheridan, for the Professor, was marked by two important events, namely: a communication from the living present, and another from the dead past. The first came, as the postmark showed, by way of Lindsey, on the Solomon river. The Professor said it was simply an answer to some scientific inquiries, but, to our intense amusement, he blushed like a school-girl when Sachem bluntly remarked that the handwriting was feminine, and that the scientific information in question must certainly be contraband,as it was not offered for our benefit at all.

A geologist in love is a phenomenon. The dusty museum is no plaçe for Cupid. In his flights, the mischievous boy is apt to hit his head against fossil lizards, and his darts are intercepted by skulls which were petrified before he ever wandered through Paradise and tried his first barb on poor Adam. The atmosphere which inwraps the geologist comes from an unlovable age, in which monstrosities existed only by virtue of their expertness in devouring other monstrosities. No stray spark of love-light flickered, even for an instant, over that waste of waters and gigantic ferns.

It was apparent that science would suffer, unless the Solomon river was included in our homeward route. We had examined the heart of Buffalo Land, having traversed its center from east to west, and our party was disposed to oblige the Professor by returning along the northern border. Southward two hundred miles was the Arkansas, flowing near the southern limit of the buffalo region. While there were some reasons why we desired to visit it, and though it was, perhaps, equally rich in game, it promised nothing of greater interest, upon the whole, than the district we now proposed traversing. But of this more in the next chapter.

Toward evening came our introduction to what we were pleased to imagine was a beauty of the past, which happened thus: As we were wandering among the Mexican teamsters loafing around the depot, an urchin, with half a shirt and very crooked legs, ran up to us, and exclaimed, over a half masticated morsel of cheese, "Mister, there's a bufferler!" His crumby fingers pointed in a direction midway between the horizon and a Mexican donkey, which its owner was trying to drag across the valley, and there, true enough, on the side of a brown ridge, not a mile off, we saw the game, feeding as usual.

Here was a chance for horseback hunting again, which we had not attempted for several days. And what a splendid opportunity of showing the natives how well we could do the thing! Our wagons had groaned under the burden of pelts and meats with which we had loaded them, and we were suffering just then from that dangerous confidence which first success is so apt to inspire.

Half the pleasure of hunting, if sportsmen would but confess it, consists in showing one's trophies to others. It was not at all surprising, therefore, that the send-off found two-thirds of our force in the field. The day was warm, and, though the hunters ran far and fast, the bison went still further and faster, and escaped. He led us, however, to greater spoil than his own tough carcass; for underneath the sod which his hoofs spurned, lay a treasure which glittered as temptingly to geological eyes as gold to the miner, when first struck by his prospecting pick.

The Professor trotted out of town with becoming dignity, following the hunters merely to avail himself of their protection, while examining the ridges around. A mile out, the heat and his rough-paced nag proved too much for him, and he threw himself upon the ground for a rest. Lying there, watching idly the little insects wandering about, his attention was attracted to a colony of burrowing ants, who, with a hole in the earth half an inch in diameter, were continually coming up, rolling before them small grains of sand and pebbles, the latter obtained far below, and a small mound of them already showing the extent of their patient labors. The Professor began to mark more closely the tiny builders, imagining that he could distinguish one of the citizens going down, and recognize him again as he came up again with his burden from below.

Occasionally, it seemed to the observant savan, something blue was brought out, which glittered more than sand. Looking closer, he discovered that the shining particles were beads of some bright substance, and resembling exactly those worn by the Indians of to-day. It thrilled him, as if he had been brought face to face with the far-off ages, when the world was young. Beneath, evidently, lay the dead of some forgotten tribe, and horse and man were resting upon a place of sepulcher. There was no mound to mark the spot, and if any ever existed, the seasons of ages had obliterated it. The savage races which now roam the plains never bury their dead, but lay the bodies on scaffolds, or hang them in trees. And so these little ants, robbing the graves far beneath us, were bringing to our gaze, on a bright summer day in the Nineteenth Century, the mysteries of ages already hoary with antiquity when Columbus first saw our shores.

We found ourselves wondering to what race the hidden dead belonged, and whether the unpictured maidens of those days were pleasant to look upon, or true ancestors of the hideous and unromantic creatures who, with their savage lords, now roam the plains. Thinking of the tribes of the past brought those of the present to mind, and, not wishing to have our hair presented as tribute to some maiden wooed by treacherous Cheyenne, we turned our horses' heads homeward, bringing the beads with us, safely deposited in one of our entomologist's pocket-cases. They remain among the trophies of our expedition, and Mr. Colon has lately written me that he will have an excavation made, during the present year, at the spot where they were found.

These beads, I can not but think, form one link in a chain connecting an ancient people, perhaps the mound-builders, with the savage tribes of the present. There is a tradition among some of the Western Indians that, centuries ago, a people, different in language and form from the red men, came from over the seas to trade beads for ponies. The buffaloes were then larger, and the climate warmer, than now. Dissensions finally arose, in which the strangers were killed. Is there not reason to believe that this tradition gives us a glimpse of the time when some of the large mammals still existed on the plains, and the genial sun looked down upon pastures clothed in rich vegetation—a time and region, probably, of perennial summer?

Once, during our stay in Kansas, we were directed by a hunter to a spot where he had seen portions of an immense skeleton, and there found one vertebra only remaining of a mastodon. It afterward transpired that, shortly before our trip, some Indians had passed Fort Dodge with the large bones lashed on their ponies, taking them to a medicine-lodge on the Arkansas, to be ground up into good medicine. They stated that the bones belonged to one of the big buffaloes which roamed over the plains during the times of their fathers. At that period, the Happy Hunting Ground was on earth, but was afterward removed beyond the clouds by the Great Spirit, to punish his children for bad conduct.

Many reasons, besides dim traditions, exist for the belief that those mysterious nations whose paths we have been able to trace from the Atlantic west, and from the Pacific east, pushed inward until they met in the middle of the continent. The numerous mounds in the Western States, with the curious weapons and vessels which they contain, show that the nations then existing, and migrating toward the interior, were not only powerful but essentially unlike our modern Indians. To instance but one illustration of this, there are near Titusville, Pa., ancient oil wells, which bear unmistakable evidences of having been dug and worked by the mound-builders. Thus they speculated in oil, which of itself is a token of high civilization.

Coming east from the Pacific coast, we find existing on the very edge of the desolate interior extensive ruins of ancient cities, of whose builders even tradition gives no account. By these and other remains which the gnawing tooth of Time has still spared to us, the people of those days tell us that they were full of commercial energy; and who knows but they may have been as determined as our nation has ever been, to push trade across from ocean to ocean? It is highly probable also that the Indians of the interior were then far superior to the present tribes, as seems very fairly determined by many of the traditions and customs which obtain among the latter.

In view of the foregoing considerations, it is not remarkable that the beads, denoting, as they did, a place and manner of burial unlike that of the savages of the plains, interested us so much. It was a leaf, we could not but think, from the lost history of the mound-builders.

A noticeable feature of life on the plains is the sod-house, there called an adobe, from some resemblance to the Mexican structures of sun-dried brick. The walls of these primitive habitations are composed of squares of buffalo-grass sod, laid tier upon tier, roots uppermost. A few poles give support for a roof, and on these some hay or small brush is laid. Then comes a foot of earth, and the covering is complete. When well-constructed, these houses are water-proof, very warm in winter, and cool in summer; but when the eaves have been made too short to protect the walls, the latter are liable to dissolve under a heavy shower. During a sudden rain at Sheridan, being obliged to turn out early one morning to protect some goods, we discovered that the neighboring habitation had resolved itself into a mound of dirt, resembling somewhat a tropical ant-hill. We were still gazing at the ruins, when the owner, clad in the brief garment of night-wear, came spluttering through the roof, like a very dirty gnome discharged by a mud-volcano. While he stood there in the rain, letting the falling flood cleanse him off, he remarked, in a manner that for such an occasion was certainly rather dry—"Lucky that houses are dirt-cheap here, stranger, for I reckon this one 's sort o' washed!"

A person of small capital, as may readily be inferred, can live very comfortably on the plains. His house may be built without nail or board, and his meat may be obtained at no other expense than the trouble of shooting it.

We saw many wooden buildings at the different stage stations, which had subterranean communications with little sod watch-towers, rising a couple of feet above the ground, at a distance of forty or fifty yards from the main building. Loop-holes through their walls afforded opportunities for firing, and if the wooden stations were burned, the occupants could find a secure retreat. We heard of but one occasion in which the tower was ever used, but then it was most effectively, the savages, gathered close around the main building, being surprised and put to sudden flight, by the murderous fire which seemed to spring out of the ground at their rear.

CHAPTER XXVII.

OUR PROGRAMME CONCLUDED—FROM SHERIDAN TO THE SOLOMON—FIERCE WINDS—A TERRIFIC STORM—SHAMUS' BLOODY APPARITION AND INDIAN WITCH—A RECONNOISSANCE—AN INDIAN BURIAL GROVE—A CONTRACTOR'S DARING AND ITS PENALTY—MORE VAGABONDIZING—JOSE AT THE LONG BOW—THE "WILD HUNTRESS'" COUNTERPART—SHAMUS TREATS US TO "CHILE"—THE RESULT.

"GENTLEMEN," said the Professor, next morning, at breakfast, "We have well-nigh exhausted Buffalo Land. North of us some twenty miles, the upper waters of the Solomon may be reached. I believe that district to be rich in fossils; it is also interesting as the path over which the red men have so often swept on their missions of murder. The valley winds eastward and southward during its course, and will discharge us at Solomon City, a point well back on our homeward journey. There our expedition may fitly disband. Should it be considered desirable, during the coming year, to explore the wild territories of the north-west, we can meet at such place as may be designated. What say you?"

Our response was a unanimous vote in favor of accepting the programme thus sketched out. Some of us desired the trip, and all knew that the Professor would go at any rate.

Our path lay over the same undulating plain that we had been traversing for many weeks, the wind blowing fiercely in our teeth. The violent movement of the air over this vast surface is often unpleasant, and during a severe winter is more dangerous than the intense cold of the far north, as it penetrates through the thickest clothing. The winter of 1871–2, when numbers of hunters and herders were frozen to death, illustrated this to a painful degree. The months of December and January are usually mild, and no precautions were taken. On the morning of the most fatal day, it was raining; in the afternoon, the wind veered and blew cold from the north, the rain changing to sleet, and this, in turn, to snow so blinding that objects became invisible at the distance of a few feet.

After the storm, near Hays City, five men belonging to a wood-train were found frozen to death. They had unloaded a portion of their wood, and endeavored to keep up a fire, but the fierce wind blew the flames out, snatching the coals from the logs, and flinging them into darkness. The men seized their stores of bacon and piled them upon fresh kindling, but even the inflammable fat was quenched almost instantly. One of another party, who finally escaped the same sad fate, by finding a deserted dugout, said it seemed as if invisible spirits seized the tongues of flame and carried them, like torches, out into the awful blackness. Thousands of Texas cattle perished during that storm. One herder, in order to save his life, cut open a dying ox, and, after removing the entrails, took his place inside the warm carcass.

We noted a curious incident, relative to the wind's fantastic freaks on the plains, while at Sheridan. One day, during the prevalence of a north wind, we observed all the old papers, cards, and other light rubbish which ornament a frontier town, moving off to the south like flocks of birds. Two days afterward, the wind changed, and the refuse all came flying back again, and passed on to the northward.

On the first evening of our homeward journey from Sheridan, we encamped on what appeared to be a small tributary of the upper Solomon. While the tents were being pitched, and the necessary provisions unloaded, Shamus strolled toward a clump of trees half a mile off, in hopes of securing a wild turkey to add to his stores. He soon came running back in a great fright, to tell us that, as he was passing among the trees, the black pacer of the plains, with its bloody master in the saddle, had started out of a bottom meadow just beyond, and fled away into the gloom. This was a sufficiently ghostly tale in itself, but it was not all; Shamus further averred that as he turned to fly, he saw a hideous Indian witch swinging to and fro in a tree directly before him. The spot was unwholesome, he assured us, and he urged instant removal.

It seemed evident that our cook had some foundation for his fears, as his terror was too great and his account too circumstantial for the matter to be simply one of an excited imagination. If there were Indians close by, it was necessary that we should know it at once, and avoid the danger of an attack at dawn. We organized a reconnoissance immediately, and, six men strong, moved toward the timber. Scattering as much as possible, that concealed savages might not have the advantage of a bunch-shot, we cautiously reached the border of the trees, and entered their shadows. We breathed more freely; if tree-fighting was to be indulged in, we now had an equal chance. It is a trying experience, reader, to advance within range of a supposed ambuscade, and the moment when one reaches the cover unharmed is a blessed one. The logs and stumps which seemed so hideous, when death was thought to be crouching behind, suddenly glow with friendship, and one is glad to know that he can hug such friends, should danger glare out from the bushes ahead.

As we walked forward, Shamus' witch suddenly appeared before us. It was the body of a papoose, fastened in a tree.

The spot was evidently an Indian burying-ground. The corpse had been loosened by the wind, and now rocked back and forth, staring at us. It was dried by the air into a shriveled deformity, rendered doubly grotesque by the beads and other articles with which it had been decked when laid away. We had neither time nor inclination to explore the grove for other bodies, preferring our supper and our blankets. As Shamus stoutly held to the story of the phantom pacer, we were forced to conclude that some stray Indian, from motives of either curiosity or reverence, had been visiting the grove when frightened out of it by our cook. In the gathering gloom, a red shirt or blanket would have answered very well for bloody garments.

These burial spots are held in high reverence by the Indians, and their hatred of the white man receives fresh fuel whenever the latter chops down the sacred trees for cord-wood. On one occasion, a contractor destroyed a burial grove, a few miles above Fort Wallace, to supply the post with fuel. The first blow of the axe had scarcely fallen upon the tree, when some Indians who chanced to be in the neighborhood sent word that the desecrator would be killed unless he desisted. Messages from the wild tribes, coming in out of the waste, telling that they were watching, ought to have been warning sufficient. But he was reckless enough to disregard them, and continued his work. The trees were felled and cut up, and the wood delivered. The contractor went to the post for his pay, and as he took it, spoke in a jocose vein of the threat which had come to naught.

Soon afterward, he set out for camp. Midway there, he heard the rush of trampling hoofs, and looking up, his horrified gaze beheld a band of painted savages sweeping down upon him from out the west. Five minutes later, he lay upon the plain a mutilated corpse, and every pocket rifled. The Indians had fulfilled their threats. The trees which to them answered the same purpose that the marble monuments which we erect over our dead do among us, had been broken up by a stranger, and sold. They acted very much as white men would have done under similar circumstances, except that the purloined greenbacks were probably scattered on the ground, or fastened, for the sake of the pictures, on wigwam walls, instead of being put out at interest.

Our little adventure gave rise to another evening of "vagabondizing." Each one of our men, including the Mexicans, had some Indian tale of thrilling interest to relate, in which he had been the hero. José, a cross-eyed child of our sister Republic, spun the principal yarns of the occasion. He had commenced outwitting Death while yet an infant, being content to remain quiet under a baker's dozen of murdered relations, that he might be rescued after the paternal hacienda had taken fire, by somebody who survived.

After a careful analysis of several thousand remarkable stories which were told to us first and last during our journey, I have deemed it wise to repeat only those which we were able to corroborate afterward. Among the latter is a narrative that was given us by the guide on this occasion, having for its text a side remark to the effect that crazy Ann, the wild huntress whom we met above Hays, was not the first lunatic who had been seen wandering upon the plains. About the close of 1867, a small body of Kiowas appeared in the vicinity of Wilson's Station, a few miles above Ellsworth, being first discovered by a young man from Salina, who was herding cattle there. They rushed suddenly upon him, and he fled on his pony toward the station, a mile away. The chief's horse alone gained on him, and the savage was just poising his spear to strike him down, when the young man turned quickly in his saddle, and discharged a pistol full at his pursuer's breast, killing him instantly. Meanwhile, the half-dozen negro soldiers at the station had been alarmed, and now ran out and commenced firing. The Indians fled in dismay, without stopping to secure their dead chieftain, who was at once scalped by the station men, and left where he fell.

Next morning the soldiers revisited the place, and found that the band had returned in the night, and removed the corpse. The negroes followed the trail for a mile or more, in order to discover the place of burial, and shortly found the chief's body lying exposed on the bank of the Smoky. It had apparently been abandoned immediately upon the discovery that the scalp had been taken, from the belief, probably, which all Indians entertain, that a warrior thus mutilated can not enter the Happy Hunting Ground. Now for the apparition in question. As the soldiers approached the spot, a white woman, in a wretched blanket, fled away. In vain they called out to her that they were friends; she neither ceased her running, nor gave them any answer. The men pursued, but the fugitive eluded them among the trees, and disappeared. A few days after, she was again seen, but once more succeeding in escaping.

It afterward transpired that, a year or so before, a white girl had been stolen from Texas, and passed into possession of one of the tribes. She lost her reason before long, and, like all the unfortunate creatures of this class among the Indians, became an object of superstition at once. One morning she was missed by her captors, and a few days later a

Mexican teamster reported having seen a strange woman, near his camp, who fled when he approached her. His description left no doubt of her identity with the missing captive. I have since conversed with some of the soldiers, then stationed at Wilson, and they assured me that the white girl was plainly visible to them on both occasions. As she was never afterward seen in the vicinity of civilization, the poor creature is believed to have perished from exposure. Possibly she was making her way to the settlements, when frightened back by the negroes, who may have resembled her late tormentors too closely to be recognized as friends.

After one has been for months passing over a country stained every-where by savage outrage, it is easy to understand how the man whose wife or sister has met the terrible fate of an Indian captive, can spend his life upon their trail, committing murder. For murder it is, when revenge, not justice, prompts the blow, and the innocent must suffer alike with the guilty.

While breakfast was preparing next morning, some fiend suggested to one of our Mexican teamsters that the Americans might like a taste of Mexico's standard dish, "chile," of which, the fellow said, he had a good supply in his wagon-chest. Shamus was consulted, and assented at once, seeming delighted with the prospects of astonishing our palates with a new sensation. Know, O reader, of an inquiring mind, that chile consists of red pepper, served as a boiling hot sauce, or stew. It is believed to have been invented by the Evil One, and immediately adopted in Mexico.

Shamus succeeded admirably in his design of concocting a sensation for us. Our alderman was *ex-officio* the epicure of the party, half of his duties as a New York city father having been to study carefully all known flavors. He always tasted new dishes, and on our behalf accepted or rejected them. When, therefore, the savory stew came before us, he experimented with a mouthful. Immediately thereafter a commotion arose in camp, and Shamus fled before the righteous wrath of Sachem.

CHAPTER XXVIII.

THE BLOCK-HOUSE ON THE SOLOMON—HOW THE OLD MAN DIED—WACONDA DA—LEGEND OF WA-BOG-AHA AND HEWGAW—SABBATH MORNING—SACHEM'S POETICAL EPITAPH—AN ALARM—BATTLE BETWEEN AN EMIGRANT AND THE INDIANS—WAS IT THE SYDNEYS?—TO THE RESCUE—AN ELK HUNT—ROCKY MOUNTAIN SHEEP—NOVEL MODE OF HUNTING TURKEYS—IN CAMP ON THE SOLOMON—A WARM WELCOME.

ON the second day we reached the Solomon, and directed our course down its valley. Shamus' face was as bright as if he was about to blow up an English prison, which, for so pronounced a Fenian, indicated a happiness of the very highest degree. It was evident that Irish Mary had hold of the other end of our cook's heart-strings, and was twitching them merrily. Cupid had indeed found us in the solitude, and, as Sachem expressed it, was "whanging away" at two of our number, at least, most remorselessly.

Two days' ride brought us to the forks of the river, where a block-house had been built a year or two before, and in which we expected to find a resident. Since its abandonment by the troops, it had been occupied by an elderly man, known as Doctor Rose, who, solitary and alone, was holding this frontier post, that, when civilization came, he might possess it as a farm. We were disappointed. The barricade was deserted, and every thing about it as silent as the grave. No curling smoke uprose among the trees, and the everlasting hills and dusky prairies stretched away on all sides in weird, wild desolation. We shook the door, and called, but found no answer. It was fastened upon the inside, and as we had no right to force it, we passed on, and encamped by the "Waconda Da," or Great Spirit Salt Spring, a few miles below.

We did not suppose that the old man we had sought was so near us. Up on a high ridge only a short distance off, his body was lying, another victim of Indian murder. Savages had been raiding through the settlements below, and thinking himself exposed, he had contrived to fasten the door of the block-house from the outside, and attempted to escape in the night. No one but the red murderers saw the old man die, and how and when they met him will never be known; but his body was found near the roadside, where the path wound over a high ridge, and within sight of the Waconda, and there it was afterward laid in its lonely sepulcher by his sorrowing family.

Down on a creek below, the savages, on the previous evening, had been sweeping off the thin line of settlements, as a broom sweeps spiders' houses from the wall. Perhaps some dark demon eye, glancing up from the crimson trail, saw the old man, bending under the weight of years, feebly trying to save the few remaining days left him, and turned pitilessly aside to hurl him into that grave which, at best, could not be far off. No struggle was visible where he fell, and it is probable that they approached him with a treacherous "How, how?" and a hand-shake, and, as he gave the grasp of friendship, struck him down, and launched him into eternity.

Waconda Da, Great Spirit Salt Spring, is among the most remarkable natural curiosities of the West, and is held in great reverence by the native tribes. It presents the appearance of a large conical mass of rock, about forty feet high, shaped like an inverted bowl, and smooth as mason-work. In the center of its upper surface, is the spring, shallow at the rim, and in the middle having a well-like opening, about twenty feet in depth. Into this pool the Indians cast their offerings, ranging from old blankets to stolen watches, thereby to appease the Great Spirit. (From his location, Sachem thought the latter must be an old salt.)

We fished with a hooked stick for some time, and were rewarded by bringing up a ragged blanket and a shattered gunstock. All around the rim of the opening were incrustations of salt, and the brackish water trickled over, and ran in little rivulets down the huge sides. At the base of the rock, a dead buffalo was fast in the mud, having died where he mired, while licking the Great Spirit's brackish altar.

As no remarkable spot in Indian land should ever be brought before the public without an accompanying legend, I shall present one, selected out of several such, which has attached itself to this. To make tourists fully appreciate a high bluff or picturesquely dangerous spot, it is absolutely essential

WACONDA DA—GREAT SPIRIT SALT SPRING.

that some fond lovers should have jumped down it, hand-in-hand, in sight of the cruel parents, who struggle up the incline, only to be rewarded by the heart-rending *finale.* This, then, is

THE LEGEND OF WACONDA.

Many moons ago—no orthodox Indian story ever commenced without this expression—a red maiden, named Hewgaw, fell in love. (And I may here be permitted to quote a theory of Alderman Sachem's, to the effect that Eve's daughters generally fall into every thing, including hysterics, mistakes, and the fashions.) Hewgaw was a chief's daughter, and encouraged a savage to sue for her hand who, having scalped but a dozen women and children, was only high private or "big soldier." Chief and lover were quickly by the ears, and the fiat went forth that Wa-bog-aha must bring four more scalps, before aspiring to the position of son-in-law. This seemed as impossible as Jason's task of old. War had existed for some time, and, as there was no chance for surprises, scalp-gathering was a harvest of danger.

There seemed no alternative but to run for it, and so, gathering her bundle, Hewgaw sallied out from the first and only story of the paternal abode, as modern young ladies, in similar emergencies, do from the third or fourth. Through the tangled masses of the forest, the red lovers departed, and just at dawn were passing by the Waconda Spring, into whose waters all good Indians throw an offering. Wa-bog-aha either forgot or did not wish to do so. Instantly the spring commenced bubbling wrathfully. So far, the Great Spirit had guided the lovers; now, he frowned. An immense column of salt water shot out of Waconda high into air, and its brackish spray dashed furiously into the faces of Wa-bog-aha and Hewgaw, and drove them back.

The saltish torrent deluged the surrounding plains—putting every thing into a pretty pickle, as may well be imagined. The ground was so soaked that the salt marshes of Western Kansas still remain to tell of it, and, a portion of the flood draining off, formed the famous "salt plains." Along the Arkansas and in the Indian Territory, the incrustations are yet found, covering thousands of acres. The Kansas River, for hours, was as brackish as the ocean, its strangely seasoned waters pouring into the Missouri, and from thence into the Mississippi. It was this, according to tradition, which caused such a violent retching by the Father of Waters, in 1811. The current flowed backward, and vessels were rocked violently—phenomena then ascribed by the materialistic white man to an earthquake.

Too late the luckless pair saw their mistake, and started for the summit of Waconda, just as the angry father put in his very unwelcome appearance. Had they avoided looking toward the spring, all, perchance, might yet have been well. Without exception, the medicine men had written it in their annals that no eye but their own must ever gaze back at Waconda, after once passing it. Tradition explains that this was to avoid semblance of regret for gifts there offered the Great Spirit. Sachem, however, is of the opinion that in giving these orders the medicine men had the gifts in their eye, and simply wished time to put them in their pockets. Hewgaw could not resist the temptation to peep. Immediately around the rock all was quiet, while without the narrow circle the descending torrents were dashed fiercely by the winds. The beasts of the plains, in countless numbers, came rushing in toward the Waconda, their forms white with coatings of salt, and probably representing the largest amount of corned meat ever gathered in one place.

All the brute eyes—knightly elk, kingly bison, and currish wolves—were turned toward the top where Wa-bog-aha and Hewgaw stood, casting their valuables, as appeasing morsels, into the hissing spring. It refused to be quieted. Suddenly, the lovers were nowhere visible, and the salt storm ceased. Nothing could be found by the afflicted father, except a tress of his daughter's hair—perhaps her chignon.

The old chief declared that, just as the end was approaching, the clouds were full of beautiful colors, and the air glittered with diamonds. The white man's science, however, coldly assumes that these appearances were only the rainbows and their reflections, playing amidst the crystal salt shower.

Sabbath morning dawned upon our camp, and according to our usual custom, we lay by for the day. At ten o'clock, the Professor read the morning service. It must have been a strange scene that we presented, while uncouth teamsters and all—our family-pew the wide valley, with its seats of stones, and logs—sat listening to the beautiful language that told how the faith of which Christianity was born was cradled in a land as primitive and desolate as that which we were traversing. There, the wild Arab hordes hovered over the deserts; here, America's savage tribes do the same over the plains.

Our priest stood near one of Nature's grandest altar pieces, "Waconda Da." Reverence from the most irreverent is secured among such scenes and solitudes. Away from his fellows, man's soul instinctively looks upward, and yearns for some power mightier than himself to which to cling. The brittle straw of Atheism snaps when called upon for support under these circumstances, and the blasphemy which was bold and loud among the haunts of men, here is hushed into silence, or even awed into reverential fear.

The Professor improved the opportunity to deliver an excellent discourse upon the wonderful evidences of God's power which geology is daily revealing. His peroration was quite flowery, and in a strain very much as follows:

"Science is yet in its infancy, and many things which seem dark to us will be clear to our descendants. Future generations will doubtless wonder at our boiler explosions, and our railroad accidents. Lightning expresses will be used only for freight, while machines navigating the air, at one hundred miles an hour, will carry the passengers. Steam, electricity, and the magnetic needle have all been open to man's appropriative genius ever since the world offered him a home, and yet he has only just now comprehended them. The future will see instruments boring thousands of feet into the earth in a day, and developing measures and mysteries which the world is not now ripe for understanding. Perhaps, the telescopes of another century may bring our descendants face to face with the life of the heavenly bodies, and give us glimpses of the inhabitants at their daily avocations. Who knows but that the beings who people other worlds in the infinite ocean of space around us, compared with which worlds our little planet is insignificant indeed, are able, by the use of more powerful instruments than any with which we are acquainted, to hold us in constant review? Our battles they may look upon as we would the conflicts of ants, and they wonder, perchance, why so quarrelsome a world is permitted to exist at all."

Next morning Sachem was up at daybreak, examining the spot where Hewgaw and Wa-bog-aha met their fate, and underwent their iridescent annihilation. His offering to their memory we found after breakfast, tacked up in a prominent position beside the spring. The inscription, evidently intended as a sort of epitaph, was written on the cover of a cracker-box, and struck me as so peculiar that I was at the pains of transcribing it among our notes. I give it to the reader for the purpose, principally, of showing the unconquerable antipathies of an alderman.

In Memoriam.

Lot's wife, you remember, looked back,
(What woman could ever refrain?)
And instantly stood in her track
A pillar of salt on the plain.

If all were thus cursed for the fault,
Who peep when forbidden to look,
The feminine pillars of salt
Could never be written in book.

Hewgaw was an Indian belle
Which no one could ring—she was fickle;
Some scores of her lovers there fell
(Where she did at last) in a pickle.

Thus salt is the only thing known
Entirely certain of keeping
Flesh of our flesh, bone of our bone,
Out of the habit of peeping.

Unless the tradition has lied,
Our maiden may claim, with good reason,
That she is a well-preserved bride,
And certainly bride of a season.

Wa-bog-aha big was a brave—
The Great Spirit salted him down:
Braves seldom get corned in the grave,
They 're oftener corned in the town.

My rhyming, you find, is saline,
Quite brackish its toning and end;
The moral—far better to pine
Than wed and get "salted," my friend.

Soon after sunrise we took our way down the river, intending to reach the Sydney farm on the following day, and there spend the necessary time in preparing our specimens for immediate shipment when we should arrive at Solomon City. The Professor made desperate efforts to appear entirely wrapped up in science, and his devotion to geology was something wonderful. Hitherto he had been inclined to urge us forward, but now he made a show of holding us back. Did he do so with a knowledge that our necessities for food and forage would be sufficient spur, and was he simply shielding his weak side from Sachem's attacks?

We had proceeded but a few miles on our journey, when the guide rode back, and reported fresh pony tracks across the road ahead of us. This was an unquestionable Indian sign, but as the trail seemed to be leading north, we took no precaution; our route was over a high divide, where ambushing was impossible.

Approaching Limestone Creek, the road wound down the face of a precipitous bluff, into the valley below. We had just commenced the descent, when the now familiar cry of "Injuns!" came back from the men in front, and following closely on the cry we heard the echoing report of firearms. We looked in the direction of the sound, and saw close to the trees an emigrant wagon, while beyond it, but at fully one hundred yards' distance, four or five Indians were riding back and forth in semi-circles, and firing pistols. The emigrant stood beside his oxen, with rifle in readiness, but apparently reserving his fire.

"That man knows his biz!" exclaimed our guide, as he urged the teams forward, that we might afford rescue. "Injuns never bump up agin a loaded gun."

A gleam of calico was visible in the wagon, and another rifle barrel, held by female hands, seemed peering out in front. The general aspect of the assailed outfit reminded us strongly of the Sydney family, and suspicion was strengthened by a very unscientific yell from the Professor, as he started off at break-neck speed down the bluff for a rescue, with no other weapon whatever in his hand than a small hammer he had just been using for breaking stones. Mr. Colon seemed equally demented, following close upon Paleozoic's heels with a bug-net. Shamus, at the moment, happened to be astride his donkey, and giving an Irish war-whoop which reached even to the scene of combat, straightway charged over the limestone ledges in a cloud of white dust. Our appearance upon the scene was a surprise to Lo. The Indians stood not upon the order of their going, but "lit out on the double-quick," as our guide expressed it, and were soon out of sight.

We found that the emigrants were named Burns, the family comprising the parents and their two children. The man stated that he had no fear of the savages. He had been twice across the plains, and made it a rule never to throw a shot away. "If they can draw your fire," said he, "the fellows will charge. But they do n't want to look into a loaded gun." Mrs. Burns had come to her husband's rescue with an expedient worthy the wife of a frontiersman. Having no gun, she pointed from under the canvass the handle of a broom. This, being woman's favorite weapon, was handled so skillfully that the savages imagined it another rifle. In our log-book she was chronicled at once as fully the equal of that revolutionary hero, who one evening made prisoner of a British officer, by crooking an American sausage into the semblance of a pistol, and presenting it at the Englishman's breast.

There were two of our party who did not rejoice as they should have done, after rendering such timely aid to the Burns family. How romantic had the rescued party only proved to be the one which was at first suspected!

Where this little scene occurred, there are homesteads now, which will soon develop into thrifty farms. The blessing of a railroad can not be long deferred. A year, a month, even a week sometimes, makes wonderful changes in Buffalo Land, when the tide of immigration is rolling forward upon it. Before the present year is ended, the beautiful valley of Limestone Creek will be teeming with civilized life, and the savage red man, there is good reason to believe, has departed from it forever.

After bidding the Burns family good-bye, we traveled without further adventure until near noon, when the guide rode back, and directed our attention to some elk, which he pointed out, some distance ahead. The bodies of the herd were hidden by a ridge, but above its brown line we could plainly see their great antlers, looking like the branches of trees, moving slowly along. There was but one method of getting near the game, and that was immediately adopted. Up the side of the sloping ridge we carefully crawled, and, reaching the summit, peeped over. Half a dozen big antlered fellows, and as many does, were feeding along the slope below. Only one of them, a splendid male, was within shooting distance at all, and even for it the range was long. The guide and Muggs fired together, breaking the poor creature's shoulder.

What a startled stare the noble animals flashed back at the crack of the rifles, and how quickly they disappeared. Their trot was perfectly grand—great, firm strokes which seemed to fairly fling the bodies onward. We had hardly time to realize having fired, when their tails bade us distant adieu. It is said that no horse can keep up with the trot of the elk. If charged upon suddenly, however, from close quarters, he is frightened into an awkward gallop, and may then be overtaken easily.

Our wounded game looked formidable, and we approached cautiously. He made several efforts to run, but each time fell forward, in plunging slides, on his nose and side, rubbing the hair from the latter, and daubing the ground with blood from his nostrils. Muggs felt free to confess that even the pampered stags of England, when perilously roused from their well-kept glens, by over-fed hunters in killing coats and boots, never presented such a picture of wild beauty and agony, colored just the least bit with danger. At this "kill" we lost our black hound. Tempted to incaution by the sight of the noble elk standing wounded and at bay, or else excited by its blood, the dog sprang forward. A chance blow of the massive horns knocked him over, and in an instant more the beast had stamped him to death.

We finished the elk by a united volley, and added him to our trophies. The horns, resting upon their tips, gave space for one of our Mexicans, five feet two in stature, to pass beneath them erect. Elk hairs are remarkably elastic. Single ones obtained from this specimen stretched by trial with the fingers, and detached from the skin so easily that the latter seemed worthless.

During the day we found and secured the remains of two saurians—one about eight and the other ten feet in length, and also the tooth of a fossil horse, quite a number of curious bubble-shaped pieces of iron pyrites, and some fine petrifactions, in the way of butternuts and fragments of trees. The soft, white limestone, mentioned more than once before in this record of our expedition, appeared along our paths in fine outcrops, and contained very perfect fossil shells.

Abe, our guide, told us that a year or two previous, during a winter of unusual severity, he had found a flock of Rocky Mountain sheep feeding near the Solomon. This was the only instance which came to our knowledge of that animal having been seen upon the plains.

We had an amusing experience, before night, with turkeys, hunting them in novel style. The birds were wild from recent pursuit, and, the instant they saw us, would leave the narrow fringe of timber, and run off into the ravines. Then would commence a ludicrous chase, each rider plying spurs, and pursuing. There went Sachem, on his long-legged purchase, the beast staggering and stumbling through ravines; and Semi also, upon Cynocephalus, whose abbreviated tail was hoisted straight in air, while at the other extremity his nose stretched well out and took in air under asthmatic protests. Rearward was the Mexican donkey, arguing the point with Dobeen whether or not to enter the race. Ahead of all went the wild turkeys, running like ostriches. The bird is a heavy one, and its short flights and runs, therefore, though rapid, can not be long continued. Seeing the pursuit gaining, it would turn to the woods again for protection. Other riders would there head it off, and soon, completely exhausted and only able to stagger along, it was easily taken. In this manner, we obtained over twenty turkeys while passing along the river.

That evening we reached the little settlement on the Solomon, which was the Canaan of all our wanderings to certain members of our party, and went into camp among the Sydneys and their neighbors. Our welcome was a warm one, and it took Shamus but a few moments to find our friend's kitchen, where he at once installed himself in the dual capacity of lover and assistant cook, discharging the duties of each position to the entire satisfaction of all concerned. Our supper with the Sydney family seemed like civilization again, notwithstanding that we were still on the uttermost bounds of civilized manners and customs. The Professor, sitting next to Miss Flora, was the very picture of happiness, and "all went merry as a marriage bell."

PRAIRIE CHICKENS.

HEAD OF AN ELK.

WILD TURKEY.

BEAVER.

MORE OF OUR SPECIMENS—PHOTOGRAPHED BY J. LEE KNIGHT, TOPEKA, KANS.

Even Sachem ceased to sulk before the meal was ended.

At dusk, as we were assuring ourselves by personal inspection that the camp was in proper order, a familiar form came stalking toward us in the gathering gloom. "Tenacious Gripe!" cried the Professor; and so it was. Our friend's ribs had been repaired, and he was now on a mission along the Solomon river, holding railroad meetings in the different counties. The progressive westerner, when he has nothing else to do, is in the habit of starting out on a tour for the purpose of inducing the dear people to vote county bonds for a new railroad, and such a westerner was Gripe.

CHAPTER XXIX.

OUR LAST NIGHT TOGETHER—THE REMARKABLE SHED-TAIL DOG—HE RESCUES HIS MISTRESS, AND BREAKS UP A MEETING—A SKETCH OF TERRITORIAL TIMES BY GRIPE—MONTGOMERY'S EXPEDITION FOR THE RESCUE OF JOHN BROWN'S COMPANIONS—SCALPED, AND CARVING HIS OWN EPITAPH—AN IRISH JACOB—"SURVIVAL OF THE FITTEST"—SACHEM'S POETICAL LETTER—POPPING THE QUESTION ON THE RUN—THE PROFESSOR'S LETTER.

SUPPER over, we made an engagement with our hospitable friends for their presence at a sort of "state dinner" we proposed giving the next day, and then returned to our own camp. A number of the settlers soon came strolling in, and among them one bringing a most remarkable dog, of the "shed-tail" variety. The animal was well known to fame in that section, for having attacked some Indians who had taken his mistress captive and were endeavoring to place her upon one of their ponies, and so delaying them that the neighbors were able to arrive and give rescue. It was claimed that thirty shots were fired at him without effect, which, if true, proved that either those Indians were exceedingly bad marksmen, or that the small fraction of caudal appendage which the beast possessed acted as a protective talisman.

We had often seen dogs without tails, but previous to this had always supposed that a depraved human taste, not nature, was at the root of it. Tail-wagging we had considered as much the born prerogative of a dog as a laugh is that of man. It is true some men do not laugh, but the child did. A dog's tail embodies his laughing faculty, or rather one might call it a canine thermometer. It rises and falls with his feelings, in moments of depression going down to zero between his legs, and again rising when the canine temperature becomes more even.

"That thar dorg, stranger, is of the shed-tail variety," said its owner, when we solicited information. "Whole litter had nothin' but stumps. Killed most on 'em off, 'cause, havin' nothin' to wag, visitin' people could n't tell whether they was goin' to bite, or be pleased. Some time ago, a travelin' school-teacher giv' him a plaguy Latin name, but we call him Shed, for short. He knows, just as well as you and I, that he 's in the wrong, latterly, and as soon as you look at him, or touch where the tail ought ter be, he hides and howls. He 's sensitive as a human."

Saying this, our new acquaintance leaned over the dog, which was lying asleep, and gave the animal what he called a "latterly touch." Although it was but the gentle contact of a finger tip, the poor creature jumped up, uttered a dismal howl, and fled off among the wagons.

"That dorg," continued the owner, "would be one of the best critters out, if it was n't for his short cut.

He 'll fight Injuns, or wild cats, and take any amount of blows on his head, if they 'll only avoid his misfortin.' "

We remarked that he seemed to have been shot in the side, some time.

"Yes, got a whole charge of quail shot slapped inter him. You see the way it was, wer this. Most every section has one or two scraggy, rattle-brained fellers, allers loungin' round, takin' free drinks, and starvin' ther families. Whar we come from was one of this sort, never of no account to no one. We had a temperance meetin' one day, and this Hib, as they called him, wer opposed to it. He was afraid they 'd shut up Old Bung's whisky shed. Well, we was all a gathered, listenin' to the serpent and its poisoned sting, and that sort o' thing, and had about concluded to go for Old Bung, when that contrairy, ornery Hib broke us up. He goes and gets a fresh coon skin, and sneaks all round the school-house, draggin' it arter him, and makin' a sort o' scented circle. Then he goes and gets Shed Tail there, who was powerful on coons, and sets him on that thar track. Shed give just one sniff, and opened right out. The way he shied round that school-house wer a sin. In five minutes, all the dogs of the village were at his heels, and goin' round that circle like the spokes in a wheel.

"It was just a round ring of the loudest yelling you ever heard. Every dog thought the one just ahead of him had the coon. All the meetin' folks come a pourin' out, with sticks and chairs, and what with beatin' and coaxin' they got all off the trail but old Shed. Half the people went to chasin' that dorg, while the balance held onto the others. But Shed just stuck to that coon track, like all possessed, dodgin' atween our legs, or sheerin' off, and catchin' ther trail agin just beyond. He finally upset Old Squire Bundy's wife, and the Squire got mad, and slapped some No. 7 into his ribs."

The shed-tail's owner, waxing more and more eloquent with his subject, had just commenced the narrative of another Indian battle in which his favorite had figured, when we became interested in a wordy political combat between Tenacious Gripe and a genuine specimen of the "reconstructed," the first and only one of that genus that we saw in Kansas. His clothes had the famous butternut dye, and his shirt bosom was mapped into numerous creeks and rivers by the brown stains of tobacco overflows. The dispute waxed warm, and grew more and more prolific of eloquence. At length, the reconstructed beat a retreat, and our orator was left in triumphant possession of the field.

Drawing fresh inspiration from his success, Gripe devoted another hour to an account of the early struggles in Kansas against these "mean whites." He gave us many vivid descriptions of the time when men died that their children might live. Among other relations was that of the expedition under Montgomery, to rescue the two companions of old John Brown from the prison at Charlestown, Virginia, a short time after the stern hero himself had there been hung.

The dozen of brave Kansas men interested in the

enterprise reached Harrisburg, with their rifles taken apart and packed in a chest, and sent scouts into Virginia and Maryland. It was the middle of winter, and deep snow covered the ground. They intended, when passing among the mountains, to bear the character of a hunting party. Every member of that little band was willing to push on to Charlestown, notwithstanding the whole State of Virginia was on the alert, and pickets were thrown out as far even as Hagerstown, Maryland. The plan was, by a bold dash to capture the jail, and then, with the rescued men, make rapidly for the seaboard. Although the expedition failed, it gave the world a glimpse of that heroic western spirit which was not only willing to do battle upon its own soil, but content to turn back and meet Death half-way when comrades were in danger.

Gripe did not accompany the expedition. Yet he grew so eloquent over the deep snow that stretched drearily before the little band, the gloomy mountains which frowned down defiance, and the people, far more inhospitable than either, who stood behind the natural barriers, filled to fanaticism with suspicion, fear, and hate, that we were sorry he had not been of the party. A man of such congressional qualifications as were his, might have been able to steal even the prisoners.

On other matters of Kansas history, Gripe could speak from personal experience. He had twice entered the territory during the period when the Free State and pro-slavery forces were doing battle for it. In one instance, the journey had been overland through Missouri, and in the other, up the Missouri River. On the first occasion, he had suffered numberless indignities at the hands of border ruffians, and would have been killed, had there been any thing in the least degree stronger than suspicion for them to act upon. On the other trip, the steamboat was stopped at Lexington, and a pro-slavery mob boarded the vessel, and searched for arms. The whole fabric of Kansas material which Gripe wove for us that evening was figured all over with battles, and murders, and tar-and-feather diversions. Had we been writing a history of the State, we might have accumulated a fair share of the material then and there.

Another subject this evening discussed around our camp-fire was the future of the vast plains which we had been traversing. Two or three of the settlers were ranchemen, who had lived in this region for many years. They were very enthusiastic about the section of their adoption, and affirmed stoutly that within fifteen years the whole tract would be under cultivation.

I can answer for our whole party that, beyond a doubt, the climate is healthy and the soil rich. For the first one hundred miles, after reaching the eastern boundary of the plains, springs and pure streams abound. Further west, the water supply is not so plentiful. On only one occasion, however, did we suffer any inconvenience from this, and that was upon the very headwaters of the Saline. Going into camp late, coffee was hastily prepared, and the quality of the water not noticed. It proved to be quite salty, and as we drank liberally of the coffee, and were unable afterward to find a spring, our sufferings before morning amounted to positive torture. Each one of the party found that his lungs were benefited by our sojourn on the plains. I believe that a consumptive could find decidedly more relief in Buffalo Land than among the mountains further west.

During the evening, we added considerably to our already very full notes concerning the wild tribes of the western plains. So many are the "true tales of the border" which one can hear in a few months of such journeyings as ours, that the recital of even a tithe of the number would become tiresome. The red-bearded owner of "Shed-tail" added to our store, by relating an adventure which he claimed had occurred to himself and Buffalo Bill, when they were teamsters together in an overland train. It was to the effect that while riding ahead of the wagons, to find a crossing over the Sandy, they discovered the skeleton of a man lying at the foot of a cottonwood tree. As they dismounted for the purpose of finding some means, if possible, of identifying the remains, their attention was caught by letters cut in the bark. These they deciphered sufficiently to see that it had been an attempt by some weak hand to carve a name. A broken knife, lying near the bones, told plainly enough who the worker at the epitaph had been, and other signs revealed to the frontiersmen the whole death history. The man had been assailed by savages, scalped, and left as dead. The work of the knife showed that he must have recovered sufficiently to crawl to the tree, and there make a faint effort to leave some record of his name and fate. The straggling gashes indicated that he had continued the task even while death was blinding his eyes. A few more drops of blood, and perhaps the mystery of years, now shrouding the history of some family hearth-stone, would have been cleared away.

We had no opportunity of verifying this story of red beard's, but as no occasion existed for telling a lie, and the neighbors of the narrator there present seemed much interested in the account, we accepted it as truth. It was apparently no attempt to impose upon the strangers. But I would here state, as a specimen feature of the frontier experience of all travelers, that whenever, at any of our camps, surrounding ranchemen or hunters discovered any member of our party taking notes, there were straightway spun out the toughest yarns which ever hung a tale and throttled truth.

Of one fact our journey thoroughly convinced us. Lo's forte has no connection with the fort of the pale-faces. An unguarded hunter, or a defenseless emigrant wagon, or unarmed railroad laborer, gratifies sufficiently his most warlike ambition. The savages of the plains, in their attacks upon the whites, have been like bees, stinging whenever opportunity offers, and immediately disappearing in space. Their excuses for the murders they commit have been as various as their moods. At one time it is a broken treaty, at another the killing of their buffalo, and trespassing upon the hunting-grounds, and again it is some other grievance. It may be some gratification for them to know that it is estimated that, until within the last three years, a white man's scalp atoned for each buffalo killed by his race.

In our various wars with the Indians, it is worthy of remark the bison have been like supply posts at convenient distances, to the hostile bands. Traveling without any supplies whatever, and therefore rapidly, a few moments suffice to kill a buffalo near the camping spot, and roast his flesh over the chips. The pony, meanwhile, makes a hearty meal on the grass. On the other hand, our troops, in pursuit of these bands, have had to encumber themselves with baggage wagons, or pack-mules, bearing food and forage.

Among our notes, I find recorded many incidents illustrative of the aptitude which the savage mind possesses for dissimulation. For instance, in our council at Hays City, White Wolf could apparently understand only our sign language; yet when the interpreter advised the Professor, in good English, not to accept the little Mexican *burro*, unless content to return its weight in something much more valuable than jackass meat, the chief could not refrain from smiling. As Indians are not given to facial revelations, the colloquy must have struck him as very apropos and very amusing. We concluded then and there, that it was unsafe to talk Indian sign with the savages for effect, and meanwhile express our real sentiments to each other in English; and upon this opinion we habitually acted thereafter.

This was our last night together as a party. The Professor had signified his intention of remaining a few days longer upon the Solomon, for the purpose of studying the surrounding country. Shamus had asked a discharge, in order to engage as farm hand for Mr. Sydney—an Irish Jacob taking to agriculture as a means of obtaining his Rachel. We received numerous invitations to divide our party for the night among the settlers, and, glad to enjoy again the luxury of a roof, Sachem and I gratefully accepted the hospitabilities of a neighboring log-cabin among the trees.

The next day was busily occupied in separating from our loads such things as the Professor and Shamus required for their further sojourn in the Solomon valley. The morning following, we bade them both good-bye, and have seen neither leader or servant since. With but one mishap, the remainder of our party reached safely the more familiar haunts of civilization. Doctor Pythagoras was the victim of our exceptional misfortune. While attempting to mount his transformed prize-fighter, the metamorphosed bully struck out from the shoulder, and the doctor was floored. We found it necessary to carry him upon a rude stretcher to Solomon City, and provide him with a section on a sleeping car for transit to the East. As we shook his hand at parting, and bade him a last good-bye, he exclaimed, "My young friends, I can not die yet. I shall recover and outlive you all. I believe in the theory of the 'survival of the fittest.'"

Ever since our return, the tide of emigration, pouring onward from the Atlantic, has lapped further and further out upon the surface of the plains; and still, as truly now as when good old Bishop Berkeley first wrote the line, "the Star of Empire westward takes its way."

While I was preparing these notes for the press, I received the following characteristic letter from Sachem, dated at his haunt in New York. It was at first a puzzle, but I found the key in a note inclosed by him, which he had lately received from the Professor.

SACHEM'S LETTER.

To crack a head and break a heart,
  Are known as Paddy's forte;
In kitchen, jail, or low-back cart—
  No matter where—he 'll court.

To don a rig, and dance a jig,
  Attend a wake or wedding,
He 'll sell his own or neighbor's pig
  And only rag of bedding.

He lives a happy, careless life,
  Hand to mouth, and heart in hand;
Ready for either love or strife,
  Building castles on the sand.

With peck of trouble ever full,
  Good measure, running over,
He deals in stock—the Irish bull,
  And with it, lives in clover.

Love's labor is the only taste
  That Paddy's mind inherits:
He thinks, where maidens run to waste,
  The harem has its merits.

And so Dobeen, upon his course,
  Love's gallop quick began;
The gal up on the other horse,
  He courted, as they ran.

The bows around the maid were more
  Than suited to her mind;
Cupid and Shamus rode before,
  The savage rode behind.

They each pursued the maiden coy,
  Two wooed her *a la* bow;
The arrow tips of one were joy,
  The other's tips were woe.

'T is said that Shamus won the race,
  And saved his hair and bacon:
If Mary loved his wooing pace,
  His heart may stop its achin'.

And this was the Professor's letter, which had evidently set the aldermanic machine to grinding doggerel again:

"ON THE SOLOMON,
LINDSEY, OTTAWA COUNTY, KANSAS.

. . . . "I have run down here after my mail. Am progressing finely with my studies. Shamus had an adventure yesterday. Mary and he rode over on horseback to a neighbor's, a mile away, and on the return were pursued by an Indian. Hard riding brought them in safely. Mary tells her mistress that, during the terrors of the chase, Shamus would not refrain from courting. He lashed her horse, and spurred his, and popped the question, alternately.

"I shall probably remain here a month or so longer, as I am much interested in the *Flora* of the Solomon Valley."

The italicized word in the last sentence is underscored, and its initial letter bears evidence of having been maliciously transformed into a capital by Sachem.

THE END.

APPENDIX.

PRELIMINARY TO THE APPENDIX.

THE officials of the new States and Territories are constantly overwhelmed with letters of inquiry from all parts of our own country and the Canadas, and even from Europe. Some of the writers wish particulars concerning the opportunities that exist for obtaining homes; others seek information as to the best points for hunting; while what to bring with them, in the way of household goods, and farming implements, or guns, dogs, etc., is the common question of nearly all.

While engaged in preparing "Buffalo Land" for the press, I published in a newspaper at Topeka a brief summary of the information then at my command upon the subjects above named. The result was the receipt of a large number of letters, asking for all sorts of details, many of which I found it impossible to answer through the mail. This fact, added to the requests of various public officers, whom I take pleasure in thus obliging, has induced me to attach an appendix to the present volume, containing a condensed statement of such matters (not elsewhere described in this work) as will assist parties westward bound, whether emigrants, sportsmen, or tourists.

The Appendix which follows is divided into three chapters. The first of these embodies information of especial interest to the immense army of home-seekers who, from every quarter, are turning their eyes eagerly and hopefully toward the free and boundless West. The second chapter is designed for the use of the sportsman, and the third furnishes very valuable and instructive details concerning the topography, resources, climate, etc., of the plains, and, more particularly, a description of the larger streams, with their contiguous valleys, which drain the vast area included within the limits of Buffalo Land.

W. E. W.

APPENDIX.

CHAPTER FIRST.

FURTHER INFORMATION FOR THE HOME-SEEKER.

24

APPENDIX.

CONTENTS OF CHAPTER FIRST.

APPENDIX.

CHAPTER FIRST.

FURTHER INFORMATION FOR THE HOME-SEEKER.

COME TO THE GREAT WEST!

THE Western States and Territories afford unexampled inducements to the surplus energy and capital of the East and Europe; and the field which they spread out so invitingly to the emigrant's choice is as wide as it is magnificent. Hundreds of millions of acres of rich land—embracing bottom and prairie, timber and running water—are open for settlement. Counties are to be populated, and towns built, all over the new States and Territories. Each of these latter is an empire in itself. Great Britain could be set down within the borders of any one of them, and yet leave room for some of the German principalities. The records of the Agricultural Bureau at Washington show that, wherever the new soil has been cultivated, both the yield per acre and the quality of the crops produced are better than in the older States. The balance of power is moving westward, and the capital of the nation, it can scarcely be doubted, must eventually come also.

There is no reason why people should starve in the great cities of this broad and heaven-favored land of ours. Business men, so often besieged and worried with applications for positions in their stores and counting-rooms, might with advantage tack up a copy of the Homestead Law by their desk, and keep a further supply on hand for distribution. Every few months some poet sings of the ill-paid seamstress in the crowded town, or some hideous murder brings to light the heroine of the garret-stitched shirt. Yet, meanwhile, at Denver City, house-girls have been getting from six to ten dollars per week, and thousands could find comfortable homes throughout Kansas, Nebraska, and Colorado, with remunerative wages. Abroad, men toil, and women work in the fields, and in one year pay out from the scanty earnings which they wring from a stingy soil more than enough to purchase one hundred and sixty acres of good land in the great and growing West.

SHOULD THERE NOT BE COMPULSORY EMIGRATION?

Except in the case of the very decrepit, or totally disabled, there can be no excuse for begging, in a country which offers every pauper a quarter-section of as rich land as the sun shines upon. I suppose the millennium will commence when laws compel the cities to drive from them the idle and vicious, and make them tillers of the soil in the wilds. Instead of brooding in the dark alleys, and breeding vice to be flung out at regular intervals upon the civilized thoroughfares, these germinators of disease and crime would be dragged forth from their purlieus and hiding-places, and disinfected in the pure atmosphere of the large prairies and grand forests. Granting that it might be a heavy burden upon their shoulders at the outset, the present generation of reformers would have the satisfaction of knowing that the sores were cleansed, and that moral and physical disease was not being propagated to suffocate their children; and even although some of the present multitude of evil-doers might not be reclaimed, most of their children certainly would be. It is more profitable to raise farmers than convicts. Instead of building jails to hold men in life-long mildew, our artisans might be building steamers and cars, to carry their products to the seaboard.

"GET A GOOD READY."

Of the immense and almost boundless tracts of Western land that invite the emigrant's choice, the larger part can be homesteaded and pre-empted, and the remainder purchased on favorable terms from the different railroads. The competition among the latter for immigration has induced low prices and superior facilities for examination.

Where a number of families are coming together, the best way, as a rule, is to select commissioners from the number, to go in advance, and spy out the land, which can be done at comparatively trifling expense. On giving satisfactory proof of their mission, such representatives are nearly always able to secure low rates of fare and freight. In this way, two or three reliable agents can select a district in which a colony may settle, and make all the necessary arrangements for its transportation, and each family save a number of dollars, which will give back compound interest in the new home.

"Get a good ready" before starting, and have your route plainly mapped out; otherwise, you will buy experience at the sacrifice of many a useful dollar. And pray that your flight be not in the winter. Come at such season as will enable you to provide at least some shelter and supplies before the inclement months come on.

Furniture and provisions can be purchased at very reasonable rates at the West, and no necessity exists, therefore, for bringing one or two car loads of broken chairs, and partially filled flour barrels. Good stock will repay transportation, but common breeds are abundant and cheap on the ground. Texas yearlings can be purchased for about six dollars per head in Kansas.

HOMESTEAD LAWS AND REGULATIONS.

The following is an epitome, by a former Register of a United States Land Office, of such laws and regulations as pertain to the securing of Government land:

The Pre-emption Act of September 4, 1841, provides, that "every person, being the head of the family, or widow, or single man over the age of twenty-one years, and being a citizen of the United States, or having filed a declaration of intention to become a citizen, as required by the naturalization laws," is authorized to enter at the Land Office one hundred and sixty acres of unappropriated Government land by complying with the requirements of said act.

It has been decided that an unmarried or single woman over the age of twenty-one years, not the head of the family, but able to meet all the requirements of the pre-emption law, has the right to claim its benefits.

Where the tract is "offered," the party must file his declaratory statements within thirty days from the date of his settlement, and within one year from the date of said settlement, must appear before the Register and Receiver, and make proof of his actual residence and cultivation of the tract, and pay for the same with cash or Military Land Warrants. When the tract has been surveyed but not offered at public sale, the claimant must file within three months from the date of settlement, and make proof and payment before the day designated in the President's Proclamation offering the land at public sale.

Should the settler, in either of the above class of cases, die before establishing his claim within the period limited by law, the title may be perfected by the executor or administrator, by making the requisite proof of settlement and cultivation, and paying the Government price; the entry to be made in the name of "the heirs" of the deceased settler.

When a person has filed his declaratory statements for one tract of land, it is not lawful for the same individual to file a second declaratory statement for another tract of land, unless the first filing was invalid in consequence of the land applied for, not being open to pre-emption, or by determination of the land against him, in case of contest, or from any other similar cause which would have prevented him from consummating a pre-emption under his declaratory statements.

Each qualified pre-empter is permitted to enter one hundred and sixty acres of either minimum or double minimum lands, subject to pre-emption, by paying the Government price, $1.25 per acre for the former class of lands, and $2.50 for the latter class.

Where a person has filed his declaratory statement for land which at the time was rated at $2.50 per acre, and the price has subsequently been reduced to $1.25 per acre, before he proves up and makes payment, he will be allowed to enter the land embraced in his declaratory statement at the last-named price, viz.: $1.25 per acre.

Final proof and payment can not be made until the party has actually resided upon the land for a period of at least six months, and made the necessary cultivation and improvements to show his good faith as an actual settler. This proof can be made by one witness.

The party who makes the first settlement in person upon a tract of public land is entitled to the right of pre-emption, provided he subsequently complies with all the requirements of the law—his right to the land commences from the date he performed the first work on the land.

When a person has filed his declaratory statement for a tract of land, and afterward relinquishes it to the Government, he forfeits his right to file again for another tract of land.

The assignment of a pre-emption right is null and void. Title to public land is not perfected until the issuance of the patent from the General Land Office, and all sales and transfers prior to the date of the patents are in violation of law.

The Act of March 27, 1854, protects the right of settlers on sections along the lines of railroads, when settlement was made prior to the withdrawal of the lands, and in such case allows the lands to be pre-empted and paid for at $1.25 per acre, by furnishing proof of inhabitancy and cultivation, as required under the Act of September 4, 1841.

The Homestead Act of May 20, 1862, provides "that any person who is the head of a family, or who has arrived at the age of twenty-one years, and is a citizen of the United States, or who shall have filed his declaration of intention to become such, as required by the naturalization laws of the United States, and who has never borne arms against the United

States Government, or given aid or comfort to its enemies, shall be entitled to enter one quarter section or less quantity of unappropriated public land."

Under this act, one hundred and sixty acres of land subject to pre-emption at $1.25 per acre, or eighty acres at $2.50 per acre, can be entered upon application, by making affidavit "that he or she is the head of a family, or is twenty-one years of age, or shall have performed service in the army and navy of the United States, and that such application is made for his or her exclusive use or benefit, and that said entry is made for the purpose of actual settlement and cultivation, and not, either directly or indirectly, for the use and benefit of any other person or persons whomsoever." On filing said affidavit, and payment of fees and commissions, the entry will be permitted.

Soldiers and sailors who have served ninety days can, however, take one hundred and sixty acres of the $2.50, or double minimum lands. In all other respects they are subject to the usual Homestead laws and regulations.

No certificate will be given, or patent issued, until the expiration of five years from the date of said entry; and if, at the expiration of such time, or at any time within two years thereafter, the person making such entry—or if he be dead, his widow; or in case of her death, his heirs or devisee; or in case of a widow making such entry, her heirs or devisee, in case of her death—shall prove by two credible witnesses that he or she has resided upon and cultivated the same for the term of five years immediately succeeding the date of filing the above affidavit, and shall make affidavit that no part of said land has been alienated, and that he has borne true allegiance to the Government of the United States; then he or she, if at that time a citizen of the United States, shall be entitled to a patent. In case of the death of both father and mother, leaving an infant child or children under twenty-one years of age, the right and fee shall inure to the benefit of said infant or children; and the executor, administrator, or guardian may, at any time after the death of the surviving parent, and in accordance with the law of the State in which such children for the time being have their domicil, sell said land for the benefit of said infants, but for no other purpose; and the purchaser shall acquire the absolute title from the Government and be entitled to a patent.

When a homestead settler has failed to commence his residence upon land so as to enable him to make a continuous residence of five years within the time (seven years) limited by law, he will be permitted, upon filing an affidavit showing a sufficient reason for his neglect to date his residence at the time he commenced such inhabitancy, and will be required to live upon the land for five years from said date, provided no adverse claim has attached to said land, and the affidavit of a settler is supported by the testimony of disinterested witnesses.

In the second section of the act of May 20, 1862, it is stipulated in regard to settlers, that in the case of the death of both father and mother, leaving an infant child, or children, under twenty-one years of age, the right and fee shall inure to the benefit of the infant child or children; and that the executor, administrator, or guardian, may sell the land for the benefit of the infant heirs, at any time within two years after the death of the surviving parent, in accordance with the law of the State. The Commissioner rules that instead of selling the land as above provided, their heirs may, if they so select, continue residence and cultivation on the land for the period required by law, and at the expiration of the time provided, a patent will be issued in their names.

In the case of the death of a homestead settler who leaves a widow and children, should the widow again marry and continue her residence and cultivation upon the land entered in the name of her first husband for the period required by law, she will be permitted to make final proof as the widow of the deceased settler, and the patent will be issued in the name of "his heirs."

When a widow, or single woman, has made a homestead entry, and thereafter marries a person who has also made a similar entry on a tract, it is ruled that the parties may select which tract they will retain for permanent residence, and will be allowed to enter the remaining tract under the eighth section of the act of May 20, 1862, on proof of inhabitance and cultivation up to date of marriage.

In the case of the death of a homestead settler, his heirs will be allowed to enter the land under the eighth section of the Homestead Act, by making proof of inhabitancy and cultivation in the same manner as provided by the second section of the act of March 3, 1853, in regard to deceased preemptors.

When at the date of application the land is $2.50 per acre, and the settler is limited to an entry of eighty acres, should the price subsequently be reduced to $1.25 per acre, the settler will not be allowed to take additional land to make up the deficiency.

The sale of a homestead claim by the settler to another is not recognized, and vests no titles or equities in the purchaser, and would be *prima facie* evidence of abandonment, and sufficient cause for cancellation of the entry.

The law allows but one homestead privilege. A settler who relinquished or abandoned his claim can not hereafter make a second entry.

When a party has made a settlement on a surveyed tract of land, and filed his pre-emption declaration thereof, he may change his filing into a homestead.

If a homestead settler does not wish to remain five years on his tract, the law permits him to pay for it with cash or military warrants, upon making proof of residence and cultivation as required in pre-emption cases. The proof is made by the affidavit of the party and the testimony of *two* credible witnesses.

There is another class of homesteads, designated as "Adjoining Farm Homesteads." In these cases, the law allows an applicant *owning* and *residing* on an original farm, to enter other land contiguous thereto, which shall not, with such farm, exceed in the aggregate 160 acres. For example, a party owning or occupying 80 acres, may enter 80 additional of $1.25, or 40 acres of $2.50 land. Or, if the applicant owns 40 acres, he may enter 120 at $1.25, or 60 at $2.50 per acre, if both classes of land should be found contiguous to his original farm. In entries of "Adjoining Farms," the settler must describe in his affidavit the tract he owns and lives upon, as his original farm. Actual residence on the tract entered as an "adjoining farm" is not required, but *bona fide* improvement and cultivation of it must be shown for five years.

The right to a tract of land under the Homestead Act, commences from the date of entry in the Land Office, and not from date of personal settlement, as in case of the pre-emption.

When a party makes an entry under the Homestead Act, and thereafter, before the expiration of five years, makes satisfactory proof of habitancy and cultivation, and pays for the tract under the 8th section of said act, it is held to be a consummation of his homestead right as the act allows, and not a pre-emption, and will be no bar to the same party acquiring a pre-emption right, provided he can legally show his right in virtue of actual settlement and cultivation on another tract, at a period subsequent to his proof and payment under the 8th section of the Homestead Act.

The 2d section of the act of May 20, 1862, declares that after making proof of settlement, cultivation, etc., "then, if the party is at that time a citizen of the United States, he shall be entitled to a patent." This, then, requires that all settlers shall be "citizens of the United States" at the time of making final proof, and they must file in the Land Office the proper evidence of that fact before a final certificate will be issued.

A party who has proved up and paid for a tract of land under the Pre-emption Act, can subsequently enter another tract of land under the Homestead Act. Or, a party who has consummated his right to a tract of land under the Homestead Act will afterward be permitted to pre-empt another tract.

A settler who desires to "relinquish his homestead must surrender his duplicate receipt, his relinquishment to the United States" being endorsed thereon; if he has lost his receipt, that fact must be stated in his relinquishment, to be signed by the settler, attested by two witnesses, and acknowledged before the register or receiver, or clerk or notary public using a seal.

When a homestead entry is contested and application is made for cancellation, the party so applying must file an affidavit setting forth the facts on which his allegations are grounded, describing the tract and giving the name of the settler. A day will then be set for hearing the evidence, giving all parties due notice of the time and place of trial. It requires the testimony of two witnesses to establish the abandonment of a homestead entry.

The notice to a settler that his claim is contested must be served by a disinterested party, and in all cases when practicable, personal service must be made upon the settler.

Another entry of the land will not be made in case of relinquishment or contest, until the cancellation is ordered by the Commissioner of the General Land Office.

When a party has made a mistake in the description of the land he desires to enter as a homestead, and desires to amend his application, he will be permitted to do so upon furnishing the testimony of two witnesses to the facts, and proving that he has made no improvements on the land described in his application, but has made valuable improvements on the land he first intended and now applies to enter.

It is important to settlers to bear in mind that it requires two witnesses to make final proof under the Homestead Act, who can testify that the settler has resided upon and cultivated the tract for five years from the date of his entry.

Patents are not issued for lands until from one to two years after date of location in the District Office. No patent will be delivered until the surrender of the duplicate receipt, unless such receipt should be lost, in which case an affidavit of the fact must be filed in the Register's Office, showing how said loss occurred, also that said certificate has never been assigned, and that the holder is the *bona fide* owner of the land, and entitled to said patent.

By a careful examination of the foregoing requirements,

settlers will be enabled to learn without a visit to the Land Office the manner in which they can secure and perfect title to public lands under the Pre-emption Act of September 5, 1841, and Homestead Act of May 20, 1862.

THE STATE OF KANSAS.

Our sojourn on the plains impressed our party with a strong belief that Kansas, at no distant day, will be one of the richest garden spots on the continent. I have more particularly described the central portion of the State, but both Northern and Southern Kansas are equally as fertile and desirable.

The United States Land Offices in Kansas are located at the following places: Topeka, Humboldt, Augusta, Salina, and Concordia. The rapidity with which Kansas is being settled may readily be inferred from the fact that 2,000,000 acres of its land were sold during one year, 1870.

In our note-book, I find the outline of a speech delivered by the Professor in Topeka, and I quote a single paragraph as fitly expressing the common sentiment of our entire number:

"Gentlemen, great as your State now is in extent of territory and natural resources, she will soon have a corresponding greatness in the means of development, and in a self-supporting population. 1870 holds in her lap and fondles the infant; 1880 will shake hands with the giant. The whole surface of your land, gentlemen, is one wild sea of beauty, ready to toss into the lap of every venturer upon it, a farm. The genius which rewards honest industry stands on the threshold of your State, with countless herds and golden sheaves, smiling ready welcome to all new-comers, of whatever creed or clime."

WHAT A FARM WILL COST.

The emigrant has already been told what it will cost him to obtain government land. If this adjoins railroad tracts, he can secure what is desired of the latter at from two to ten dollars per acre.

The expense of fencing material might be fairly estimated at from twenty to thirty dollars per thousand feet for boards, and ten to fifteen dollars per hundred for posts. This is supposing that all the material is purchased. If fortunate enough to have timber on his claim, the emigrant, of course, can inclose the farm at the cost of his own labor.

I have seen many new-comers protect their fields by simply digging around them a narrow, deep trench, and throwing the earth on the inside line so as to raise an embankment along that side two feet in height. One single wire stretched along this, and secured at proper intervals by small stakes, appears to answer quite well as a cattle guard.

Osage orange grows rapidly, and is cheap, and a permanent fence can be made with it, at small expense, in the course of three or four years.

The usual cost of breaking prairie is from two to four dollars per acre. With a yoke or two of good oxen, however, this item can also be saved.

The second year the farmer can set out with safety his trees and vines, and the third or fourth year he may be considered fairly on the road to prosperity.

Laborers' wages are from twenty to thirty dollars per month and board.

I estimate that a fair statement of the prices for stock would be about as follows: Work oxen, seventy-five to one hundred dollars per yoke; cows, twenty to fifty dollars each; horses, seventy-five to one hundred and fifty dollars.

A FEW MORE PRACTICAL SUGGESTIONS.

I would say to the emigrant, Do not be influenced to select any one particular State or locality until you have more authority for the step than a single publication. Examine carefully, make up your mind deliberately, and then move with determination. It will require no very great exertion to secure a half dozen glowing advertisements from as many new Western States and Territories. It will need but little more effort to obtain from five to fifty "rosy" circulars from as many different districts in each of the separate "garden spots." After examining these until ready to sing,—

> "How happy could I be with either
> Were t' other dear charmer away,"

take down your map, and let the railroads and streams assist your choice. You have then secured yourself against one danger of the journey—that of having these same circulars flung into your lap *en route*, and being diverted by them into dubious ways and needless expenditures. But be careful, reader, that you select not as accurate beyond the possibility of a mistake the maps accompanying the circulars; otherwise, you may find yourself unable to choose between several thousand railroad centers from which broad gauges radiate like the spokes in a wheel, and your ignorance of modern geography may be brought painfully home by discovering navigable rivers where you had supposed only creeks existed. In these matters, as in every thing else connected with your "new departure," consult *all* the various sources of information within your reach.

APPENDIX.

CHAPTER SECOND.

FURTHER INFORMATION FOR THE SPORTSMAN.

APPENDIX.

CONTENTS OF CHAPTER SECOND.

CHAPTER SECOND.

FURTHER INFORMATION FOR THE SPORTSMAN.

HUNTING THE BUFFALO.

THE first matter to be determined, in planning any sporting trip, is the best point at which to seek for game. If the object of pursuit be buffalo, I should say, Deposit yourself as soon as possible on the plains of Western Kansas.* Take the Kansas Pacific Railway at the State line, and you can readily find out from the conductors at what point the buffalo chance then to be most numerous. There are a dozen stations after passing Ellsworth equally good. One month, the bison may be numerous along the eastern portion of the plains; a month later, the herds will be found perhaps sixty or eighty miles further west. As one has at least a day's ride, after entering Kansas, before penetrating into the solitude of Buffalo Land, there is ample time to decide upon a stopping place. Russell as an eastern, and Buffalo Station as a western point, will be found good basis for operations. In the former, some hotel accommodations exist; in the latter, there are several dug-outs, and hunters who can be obtained for guides.

Those who can spend a week or more on the grounds, and wish to enjoy the sport in its only legitimate way, namely,

* During the present year, the A. T. & Santa Fe R. R. will probably be finished to the big bend of the Arkansas, which will place the sportsman in one of the finest game regions of the continent.

horseback hunting, should stop at the point where they may best procure mounts, even if it necessitate a journey in the saddle of twenty miles. Ellsworth, Russell, and Hays City are the places where such outfits may generally be obtained.

For shooting bison, the hunter should come prepared with some other weapon than a squirrel rifle or double barreled shot gun. I have known several instances in which persons appeared on the ground armed with ancient smooth-bores or fowling-pieces; and in one of these cases the object of attack, after receiving a bombardment of several minutes' duration, tossed the squirrel hunter and injured him severely. A breech-loading rifle, with a magazine holding several cartridges, is by far the best weapon. In my own experience I became very fond of a carbine combining the Henry and King patents. It weighed but seven and one-half pounds, and could be fired rapidly twelve times without replenishing the magazine. Hung by a strap to the shoulder, this weapon can be dropped across the saddle in front, and held there very firmly by a slight pressure of the body. The rider may then draw his holster revolvers in succession, and after using them, have left a carbine reserve for any emergency. Twenty-four shots can thus be exhausted before reloading, and, with a little practice, the magazine of the gun may be refilled without checking the horse. So light is this Henry and King weapon that I have often held it out with one hand like a pistol, and fired.

When a herd of buffalo is discovered, the direction of the wind should be carefully ascertained. The taint of the hunter is detected at a long distance, and the bison accepts the evidence of his nose more readily than even that of his eyes. This delicacy of smell, however, is becoming either more blunted or less heeded than formerly, owing probably to the passage over the plains of the crowded passenger cars, which keep the air constantly impregnated for long distances.

Having satisfied himself in regard to the wind, the sportsman should take advantage of the ravines and slight depressions, which every-where abound on the plains, and approach as near the herd as possible. If mounted, let him gain every obtainable inch before making the charge. It is an egregious blunder to go dashing over the prairie for half a mile or so, in full view of the game, and thus give it the advantage of a long start. When this is done, unless your animal is a superior one, he will be winded and left behind.

In most cases, careful planning will place one within a couple of hundred yards of the bison. Be sure that every weapon is ready for the hand, and then charge. Put your horse to full speed as soon as practicable. Place him beside the buffalo, and he can easily keep there; whereas, if you nurse his pace at the first, and make it a stern chase, both your animal and yourself, should you have the rare luck of catching up at all, will be jaded completely before doing so. In shooting from the saddle, be very careful between shots, and keep the muzzle of the weapon in some other direction than your horse or your feet. A sudden jolt, or a nervous finger, often causes a premature discharge. In taking aim, draw your bead well forward on the buffalo—if possible, a little behind the fore-shoulder. The vital organs being situated there, a ranging shot will hit some of them, on one side or the other. Back of the ribs, the buffalo will receive a dozen balls without being checked. A discharge of bullets into the hind-quarters, is worse than useless.

While trying in the most enjoyable and practical manner to kill the game, it is very necessary to escape, if possible, any injury to yourself or horse. The Frenchman's remark on tiger hunting is very apropos. "Ven ze Frenchman hunt ze tiger, it fine sport; but ven ze tiger hunt ze Frenchman, it is not so." Care should be taken to have the horse perfectly under control, when the bison stands at bay. Unless experienced in bull fighting, he does not appreciate the danger, and a sudden charge has often resulted in disembowelment.

Never dismount to approach the buffalo, unless certain that he is crippled so as to prevent rising. One that is apparently wounded unto death will often get upon his feet nimbly, and prove an ugly customer. I knew a soldier killed at Hays City in this manner—thrown several feet into the air, and fearfully torn. Recently near Cayote Station, on the Kansas Pacific Railway, a buffalo was shot from the train, and the cars were stopped to secure the meat, and gratify the passengers. One of the latter, a stout Englishman, ran ahead of his fellows, and shook his fist in the face of the prostrate bison. The American bull did not brook such an insult from the English one, and Johnny received a terrible blow while attempting to escape. He was badly injured, and, when I saw him some time afterward, could only move on crutches.

Should the hunter on foot ever have to stand a charge, let him fire at what is visible of the back, above the lowered head, or, should he be able to catch a glimpse of the fore-shoulder, let him direct his bullet there. The bone seems to be broken readily by a ball. Against the frontal bone of the bison's skull, the lead falls harmless. To test this fully, with California Bill as a companion, I once approached a buffalo which stood wounded in a ravine. We took position upon the hillside, knowing that he could not readily charge up it, at a distance of only fifteen yards. I fired three shots from the

Henry weapon full against the forehead, causing no other result than some angry head-shaking. I then took Bill's Spencer carbine, and fired twice with it. At each shot the bull sank partly to his knees, but immediately recovered again. I afterward examined the skull, and could detect no fracture.

A person dismounted by accident or imprudence, and charged upon, can avoid the blow by waiting until the horns are within a few feet of him, and then jumping quickly on one side. After the buffalo has passed, let the brief period of time before he has checked his rush, be employed in traversing as much prairie, on the back track, as possible, and the chances are that no pursuit will be made. Should a foot trip, or a fall from the horse give no time for such tactics, then let the hunter hug Mother Earth as tight as may be. The probabilities are that the bull can not pick the body up with his horns. I have known a hunter to escape by throwing himself in the slight hollow of a trail, and thus baffling all attempts to hook him.

Accidents are rare in bison hunting, however, and the reader should not be deterred from noble sport by the mere possibility of mishaps. I have given the above advice, feeling that I shall be well repaid if it saves the life or limbs of one man out of the thousands who may be exposed. A glimpse of surgeon's instruments should not make the soldier a coward. Comparatively few people are killed by electricity, and yet lightning-rods are very popular.

The hunter who has no love for the saddle, and prefers stalking, should provide himself with some breech-loading rifle or carbine, carrying a heavy ball—the heavier the better. The most effective weapon is the needle-gun used in the army, having a bore the size of the old Springfield musket, and a ball to correspond. A bullet from this weapon usually proves fatal. But there is little genuine sport in such practice. Stalking holds the same relation to horseback hunting that "hand line" fishing does to that with the rod and reel, the fly and the spoon, or that killing birds on the ground does to wing-shooting.

In selecting from the herd a single individual for attack, the hunter should do so with some reference to the intended use of the game. For furnishing trophies of the chase, such as horns and robe, the bull will do well; but if the meat is for use, it will be advisable to sacrifice some sport, and obtain a cow or calf. I have known many an ancient bison, with scarcely enough meat on his bones to hold the bullets, killed by amateurs, and the leather-like quarters shipped to eastern friends as rare delicacies!

ANTELOPE HUNTING.

Antelope hunting is a sport requiring more strategy and caution than the one we have described. The creature is timid and swift, and inclined to feed on ridges or level lands, where stalking is difficult. Its eyes and ears are wonderfully quick in detecting danger, and the animal at once seeks points which command the surroundings. If unable to keep in view the object of alarm, immediate flight results.

The modes of hunting this game are two. If no possibility of stalking exists, a red flag may be attached to a small stick, and planted in front of the ravine or other place of concealment. The antelope at once becomes curious, and begins circling toward it, each moment approaching a little nearer, until finally within shooting distance. The other method is by careful stalking. If the animal is on a high ridge, the sides of which round upward a little, the hunter may crawl on his hands and knees until he sees, just visible above the grass, the tips of the horns or ears. Then let him rise on one knee, with gun to shoulder, and take quick aim well forward, as the body comes into view. The approach can not be too cautious, as the antelope stops feeding every minute or so, to lift its head high, and gaze around. Thus the incautious hunter may be brought, on the instant, into full relief, and the quick bound which follows discovery, rob him of the fruit of long crawling.

Rare enjoyment might be obtained by any one who would take with him, to the plains, a good greyhound. Mounted on a reliable horse, the sportsman could follow the dog in its pursuit of antelope, and be in at the death.

ELK HUNTING.

Elk must be hunted by stalking, as he speedily distances any horse. The animal is found in abundance along the upper waters of the Republican, Solomon, and Saline. I prefer its meat to that of either the buffalo or antelope. The horns of a fine male form a pleasing trophy to look at, when the hunter's joints have been stiffened by rheumatism or age.

TURKEY HUNTING.

Wild turkeys exist in great numbers along the creeks, over the whole western half of Kansas, and, where they have never been hunted, are so tame as to afford but little sport. Cunning is their natural instinct, however, and at once comes to the rescue, when needed. After a few have been shot, the remainder will leave the narrow skirt of creek timber instantly, and escape among the ravines by fast running, defying any pursuit except in the saddle. Even then if they can get out of sight for a moment, they will often escape. While the rider is pressing forward in the direction a tired turkey was last seen, the bird will hide and let him pass; or, turning the instant it is hidden by the brow of the ravine, it will take a backward course, passing, if necessary, close to the horse. As another illustration of the wily habits of the turkey, let the hunter select a creek along which there has been no previous shooting done, and kill turkeys at early morning on roosts, and the next night the gangs will remain out among the "breaks."

For this shooting, a shot-gun is, of course, the best, although I have had fine sport among the birds with the rifle. When using shot at one on the wing, the hunter must not conclude his aim was bad, if no immediate effect is observed. The flying turkey will not shrink, as the prairie-chicken does, when receiving and carrying off lead. I have frequently heard shot rattle upon a gobbler's stout feathers without any apparent effect, and found him afterward, fluttering helpless, a mile away.

GENERAL REMARKS.

The western field open to sportsmen is a grand one. Kansas, Colorado, Nebraska, Dakotah, and Wyoming, are all overflowing with game. The climate of each is very healthy, and especially favorable for those affected with pulmonary complaints. A year or two passed in their pure air, with the excitement of exploration or adventure superadded, would put more fresh blood into feeble bodies than all the watering-places in existence. Let the dyspeptic seek his hunting camp at evening, and, my word for it, he will find the sweet savor of his boyhood's appetite resting over all the dishes. After the meal, with his feet to the fire, he can have diversion in the way of either comedy or tragedy, or both, by listening to frontier tales. When bed-time comes, he will barely have time to roll under the blankets, before sweet sleep closes his eyes, and the twinkling stars look down upon a being over whom the angel of health is again hovering.

No extensive preparation for a western sporting trip is needed, as an outfit can be obtained at any of the larger towns, in either Kansas, Nebraska, or Colorado.

Of the three districts just named, I decidedly prefer the former for the pursuit of such game as I have endeavored to describe in Buffalo Land. The eastern half of Kansas furnishes chicken and quail shooting. The birds have increased rapidly during late years, and at any point fifty miles west of the eastern line, the sportsman will find plenty of work for a dog and gun. The ground lies well for good shooting, being a gently rolling prairie, with plenty of watering-places. The cover is excellent, and with a good dog there is little trouble, between August and November, in flushing the chickens singly, and getting an excellent record out of any covey.

Wild fowl shooting is poor, there being no lakes or feeding-grounds. The best sport of that kind I ever had was in Wisconsin and Minnesota.

WHAT TO DO, IF LOST ON THE PLAINS.

There have been several instances in which gentlemen, led away from their party in the excitement of the chase, when wishing to return, suddenly found themselves lost. Judge

Corwin, of Urbana, Ohio, separated in this manner from his party, wandered for two days on the plains south of Hays City, subsisting on a little corn which had been dropped by some passing wagon. He was found, utterly exhausted, by California Bill, just as a severe snow-storm had set in. Persons thus lost should remember that buffalo trails run north and south, and the Pacific Railroads east and west. It will be easy to call to mind on which side it was that the party left the road in starting out, and it then becomes a simple matter to regain the rails, and follow them to the first station.

THE NEW FIELD FOR SPORTSMEN.

South of Kansas is the Indian Territory, which probably has within it a larger amount of game than any spot of similar size on our continent. It fairly swarms with wild beasts and birds. At sunset one may see hundreds of turkeys gathering to their roosts. Buffalo, elk, antelope, and deer of several varieties, may be found and hunted to the heart's content. Within the next two years this territory will be the paradise of all sportsmen. It can now be reached by wagoning fifty miles or so beyond the terminus of the A. T. & Santa Fé Railroad. But the savage, hostile and treacherous, stands at the entrance of this fair land and forbids further advance. While there is good hunting, there is also a disagreeable probability of being hunted. Many of the tribes which formerly roamed all over the plains are now gathered in the Indian Territory. Jealous of their rights, they are apt to repay intrusion upon them with death.

The white kills for sport alone the game which is the entire support of the savage. I have often stood among the rotting carcasses of hundreds of buffaloes, and seen the beautiful skins decaying, and tons of richest meat feeding flies and maggots; and, standing there, I have felt but little surprise that the savage should consider such wanton destruction worthy of death. In the States, game is protected at least during the breeding season; but no period of the year is sacred from the spirit of slaughter which holds high revel in Buffalo Land.

It is manifest, however, that over the Indian Territory history will soon repeat itself. Railroads are pushing steadily forward; 1872 is already seeing the beginning of the end. The savage must flee still further westward, and the valleys and prairies which he is now jealously protecting will be invaded first by the sportsman, and then by the farmer. Perhaps, before that time, Congress may have taken the matter in hand, and passed laws which will have saved the noblest of our game from at least immediate extinction.

APPENDIX.

CHAPTER THIRD.

ADDITIONAL FACTS CONCERNING THE NATURAL FEATURES, RESOURCES, ETC., OF THE GREAT PLAINS AND CONTIGUOUS TERRITORY.

APPENDIX.

CONTENTS OF CHAPTER THIRD.

CHAPTER III.

ADDITIONAL FACTS CONCERNING THE NATURAL FEATURES OF THE GREAT PLAINS; THEIR PRINCIPAL RIVERS AND VALLEYS; THEIR CLIMATE, ETC., ETC.

"BY THE MOUTH OF TWO OR THREE WITNESSES."

IN my endeavors to place Buffalo Land before the public in its true light, I have felt a desire, as earnest as it is natural, that my readers should feel that the subject has been justly treated. The opinions of any one individual are liable to be formed too hastily, and the country which before one traveler stretches away bright and beautiful, may appear full of gloomy features to another, who views it under different circumstances. A late dinner and a sour stomach, before now, have had more to do with an unfavorable opinion concerning a new town or country than any actual demerits. No two pairs of spectacles have precisely the same power, and defects ofttimes exist in the glass, rather than the vision.

These considerations have been brought to my mind with especial force when, after giving an account of our own expedition, I have searched through the records of others. A portion of the descriptions which I have been able to find are the mature productions of travelers who, perched upon the top of a stage-coach, or snugly nestled inside, have undertaken to write a history of the country while rattling through it at the best rate of speed ever attained by the "Overland Mail." What the writers of this class lack in proper acquaintance with their subject they usually make up by an air of profoundness, and positiveness in expression, and the result has more than once been the foisting upon the public of a species of exaggeration and absurdity which Baron Munchausen himself could scarcely excel.

As a rather curious illustration of the numerous absurdities which have obtained currency concerning the plains, may be mentioned the statement published more than once during the winter of 1871–2, to the effect that the snow of that region is different in character from that which falls elsewhere. In support of this assumption, the fact is adduced that snow-plows sometimes have but little effect upon it, on account of its peculiar hardness, being pushed upon it, instead of through it. A little more careful examination, however, would have discovered that the snow itself is essentially similar to that which descends elsewhere, but that the wind which drives it into the "cuts" and ravines also carries with it a large amount of sand and surface dirt; and this, packing with the snow, causes the firmness in question.

The valuable surveys being made from time to time under the auspices of the Government, in charge of persons of experience and sagacity, are doing much to replace this superficial knowledge with a more correct comprehension of what the plains really are; and, altogether, we may well hope that the time is not far distant when this whole wonderful region will be as well understood as any portion of the national domain.

As the object of this work is to place before its readers all the essential information now obtainable concerning the great plains, no apology will be necessary for adding some of the observations and opinions of other competent writers upon the same subject. By far the most valuable source which I have found to draw from in this connection, is the comprehensive report published by Government, and bearing the title of "United States Geological Survey of Wyoming and Contiguous Territory, 1870. Hayden."

THE GREAT WEST.

Prof. Thomas informs us, in his report (embodied in Hayden's survey), that, lying east of the divide, "the broad belt of country situated between the 99th and 104th meridians, and reaching from the Big Horn Mountains on the north to the Llano Estacado on the south, contains one hundred and fifty thousand square miles. If but one-fifth of it could be brought under culture and made productive, this alone, when fully improved, would add $400,000,000 to the aggregate value of the lands of the nation. And, taking the lowest estimate of the cash value of the crops of 1869 per acre, it would give an addition of more than $200,000,000 per annum to the aggregate value of our products."

"One single view from a slightly elevated point often embraces a territory equal to one of the smaller States, taking in at one sweep millions of acres. Eastern Colorado and Eastern Wyoming each contains as much land sufficiently level for cultivation as the entire cultivated area of Egypt."

FALL OF THE RIVERS.

The fall of the principal rivers traversing the region above named is about as follows: Arkansas, to the 99th meridian, eleven to fifteen feet to the mile; the Canadian, the same; the South Platte, from Denver to North Platte, ten feet to the mile; the North Platte, to Fort Fetterman, seven feet to the mile. The descent of the country from Denver Junction to Fort Hays is nine feet to the mile. Thus it will be seen that abundant fall is obtainable to irrigate all the lands adjacent.

THE PRINCIPAL RIVERS AND VALLEYS OF BUFFALO LAND.

The Platte (or Nebraska), the Solomon, the Smoky Hill, and the Arkansas, are the four largest rivers of Buffalo Land proper, and form natural avenues to the eastward from the mountains which shut it in upon the west.

THE VALLEY OF THE PLATTE.

Describing this, Hayden says: "West of the mouth of the Elk Horn River, the valley of the Platte expands widely. The hills on either side are quite low, rounded, and clothed with a thick carpet of grass. But we shall look in vain for any large natural groves of forest trees, there being only a very narrow fringe of willows or cottonwoods along the little streams. The Elk Horn rises far to the north-west in the prairie near the Niobrara, and flows for a distance of nearly two hundred miles through some of the most fertile and beautiful lands in Nebraska. Each of its more important branches, as Maple, Pebble, and Logan Creeks, has carved out for itself broad, finely-rounded valleys, so that every acre may be brought under the highest state of cultivation.

"The great need here will be timber for fuel and other economical purposes, and also rock material for building. Still the resources of this region are so vast that the enterprising settler will devise plans to remedy all these deficiencies. He will plant trees, and thus raise his own forests and improve his lands in accordance with his wants and necessities.

"These valleys have always been the favorite places of abode for numerous tribes of Indians from time immemorial, and the sites of their old villages are still to be seen in many localities. The buffalo, deer, elk, antelope, and other kinds of wild game, swarmed here in the greatest numbers, and, as they recede farther to the westward into the more arid and barren plains beyond the reach of civilization, the wild nomadic Indian is obliged to follow. One may travel for days in this region and not find a stone large enough to toss at a bird, and very seldom a bush sufficient in size to furnish a cane."

THE SOLOMON AND SMOKY HILL RIVERS.

The Solomon and Smoky Hill Rivers, while possessing some of the general characteristics of the Platte, have more timber, and the entire surrounding country is uniformly rolling. The Smoky Hill is a visible stream only after reaching the vicinity of Pond Creek, near Fort Wallace. Above that point a desolate bed of sand hides the water flowing beneath. We have spoken fully of these sections elsewhere.

THE ARKANSAS RIVER AND ITS TRIBUTARIES.

The Arkansas, passing through the southern portion of the plains, has wide, rich bottoms, with a more sandy soil than is found on the streams north. Its small tributaries have considerable timber. All these valleys are being settled rapidly.

Again consulting Prof. Thomas' report, we find that "the Arkansas River, rising a little north-west of South Park, runs south-east to Poncho Pass, where, turning a little more toward the east, it passes through a canyon for about forty miles, emerging upon the open country at Canyon City. From this point to the Eastern boundary of the Territory it runs almost directly east.

"The mountain valley has an elevation of between seven and eight thousand feet above the sea, while that of the plain country lying east of the range varies from six thousand near the base of the mountains to about three thousand five hundred feet at the eastern boundary of the Territory. From Denver to Fort Hays, a distance of three hundred and forty-seven miles, the fall is three thousand two hundred and seven feet, or a little over nine feet to the mile.

"The Arkansas River, from the mouth of the Apishpa to the mouth of the Pawnee, a distance of two hundred and six miles, has the remarkable fall of two thousand four hundred and eight feet, or more than eleven feet to the mile.

"The headwaters of the Arkansas are in an oval park, situated directly west of the South Park. The altitude of this basin is probably between eight and nine thousand feet above the level of the sea; the length is about fifty miles from north to south, and twenty or thirty miles in width at the middle or widest point. At the lower or southern end an attempt has been made to cultivate the soil, which bids fair to prove a success. Around the Twin Lakes, at the extreme point, oats, wheat, barley, potatoes, and turnips have been raised, yielding very fair crops. Below this basin the river, for twenty miles, passes through a narrow canyon, along which, with considerable difficulty, a road has been made. Emerging from this, it enters the 'Upper Arkansas Valley' proper, which is a widening of the bottom lands from two to six or eight miles. This valley is some forty or fifty miles in length, and very fertile.

"The principal tributaries of the Arkansas that flow in from the south, east of the mountains, are Hardscrabble and Greenhorn Creeks (the St. Charles is a branch of the latter), Huerbano River, which has a large tributary named Cuchara; Apishpa River, Timpas Creek, and Purgatory River. On the north side, Fountain Gui Bouille River and Squirrel Creek are the principal streams affording water.

"This entire district affords broad and extensive grazing fields for cattle and sheep, and quite a number of herders and stock-raisers are beginning already to spread out their flocks and herds over these broad areas of rich and nutritious grasses. One of the finest meadows, of moderate extent, that I saw in the Territory, was on the divide near the head of Monument Creek, and near by was a large pond of cool, clear water. The temperature of this section is somewhat similar to that of Northern Missouri, and all the products grown there can be raised here, some with a heavier yield and of a finer quality, as wheat, oats, etc., while others, as corn, yield less, and are inferior in quality."

As we descend the Arkansas, the valley becomes broader, and it is often difficult to tell where the bottom ceases and the prairie commences.

This stream attracted such a large portion of the immigration of 1871 that it is already settled upon for some distance above Fort Zarah. The soil is very rich, the climate pleasant and healthy, and good success attends both stock and crop-raising.

STOCK-RAISING IN THE GREAT WEST.

Mr. W. N. Byers, who has lived for many years in Colorado, lately contributed the following valuable article to the *Rocky Mountain News*, treating more particularly of the western half of the plains:

"After the mining interest, which must always take rank as the first productive industry in the mountain territories of the West, stock-raising will doubtless continue next in importance. The peculiarities of climate and soil adapt the grass-covered country west of the ninety-eighth degree of longitude especially to the growth and highest perfection of horses, cattle, and sheep. The earliest civilized explorers found the plains densely populated with buffalo, elk, deer, and antelope, their numbers exceeding computation. Great nations of Indians subsisted almost entirely by the fruits of the chase, but, with the rude weapons used, were incapable of diminishing their numbers. With the advent of the white man and the introduction of fire-arms, and to supply the demands of commerce, these wild cattle have been slaughtered by the million, until their range, once six hundred miles wide from east to west, and extending more than two thousand miles north and south, over which they moved in solid columns, darkening the plains, has been diminished to an irregular belt, a hundred and fifty miles wide, in which only scattering herds can be found, and they seldom numbering ten thousand animals.

"There is no reason why domestic cattle may not take their place. The climate, soil, and vegetation are as well adapted to the tame as to the wild. The latter lived and thrived the year round all the way up to latitude fifty degrees north. Twenty years' experience proves that the former do equally well upon the same range, and with the same lack of care. Time, the settlement of the country, the growing wants of agriculture, the encroachment of tilled fields, will gradually narrow the range, as did semi-civilization that of the buffalo—first from the Mississippi Valley westward, where that process is already seen, and then from the Rocky Mountains toward the east; but as yet the range is practically unlimited, and for many years to come there will be room to fatten beeves to feed the world.

"This great pasture land covers Western Texas, Indian Territory, Kansas, Nebraska, and Dakota, Eastern New Mexico, Colorado, Wyoming, and Montana, and extends far into British America. The southerly and south-easterly portions produce the largest growth of grass, but it lacks the nutritious qualities of that covering the higher and drier lands farther north and west. Rank-growing and bottom-land grasses contain mostly water: they remain green until killed by frost, when their substance flows back to the root, or is destroyed by the action of the elements. The dwarf grass of the higher plains makes but a small growth, but makes that very quickly in the early spring, and then, as the rains diminish and the summer heat increases, it dies and cures into hay where it stands; the seed even, in which it is very prolific, remains upon the stalk, and, though very minute, is exceedingly nutritious.

"In so far as the relative advantages of different portions of this wide region may be thought by many to preponderate over one another, we do not appreciate them at all, but would as soon risk a herd in the valley of the Upper Missouri, the Yellowstone, or the Saskachewan, as along the Arkansas, the Canadian, or Red River. If any difference, the grass is better north than south. One year the winter may be more severe in the extreme north; the next it may be equally so in the south; and the third it may be most inclement midway between the two extremes; or, what is more common, the severe storms and heavy snows may follow irregular streaks across the country at various points. There are local causes and effects to be considered, such as permanently affect certain localities favorably or the contrary. For instance, nearer the western border of the plains there is less high wind, because the lofty mountain ranges form a shelter or wind breaker. Of local advantages, detached ranges of mountains, hills, or broken land, timber, brush, and deep ravines or stream-beds are the most important in furnishing shelter, and, as a general thing, better and always more varied pasture ground.

"There is never rain upon the middle and northern plains during the winter months. When snow comes it is always dry, and never freezes to stock. The reverse is the case in the Northern and Middle States, where winter storms often begin with rain, which is followed by snow, and conclude with piercing wind and exceeding cold. Stock men can readily appreciate the effect of such weather upon stock exposed to its influence.

"The soil of the plains is very much the same every-where. To a casual observer it looks sterile and unpromising, but, when turned by the plow or spade, is found very fertile. Near the mountains it is filled with coarse rock particles, and under the action of the elements these become disproportionately prominent on the surface. Receding from the mountains, it becomes gradually finer, until gravel and bits of broken stone are no longer seen. Being made up from the wash and wearing away of the mountains, alkaline earths enter largely into its composition, supplying inexhaustible quantities of those properties which the eastern farmer can secure only by the application of plaster, lime, and like manures. These make the rich, nutritious grasses upon which cattle thrive so remarkably, and to the constant wonder of new-comers, who can not reconcile the idea of such comparatively bare and barren-looking plains with the fat cattle that roam over them.

"Besides the plains, there is a vast extent of pasture-lands in the mountains. Wherever there is soil enough to support vegetation, grass is found in abundance, to a line far above the limit of timber growth, and almost to the crest of the snowy range. These high pastures, however, are suitable only for summer and autumn range; but in portions of the great parks and large valleys, most parts of which lie below eight thousand feet altitude above the sea, cattle, horses, and sheep live and thrive the year round. The cost of raising a steer to the age of five years, when he is at a prime age for market, is believed to be about seven dollars and a half, or one dollar and a half per year. A number of estimates given us by stock men, running through several years, place the average at about that figure. That contemplates a herd of four hundred or more. Smaller lots of cattle will generally cost relatively more. The items of expense are herding, branding, and salt—nothing for feed."

THE CATTLE-HIVE OF NORTH AMERICA.

In this connection we may very properly quote from the same writer the following paragraph in regard to the source from whence all the cattle are now brought—that great natural breeding ground, the prairie land of Texas.

" Texas is truly the cattle-hive of North America. While New York, with her 4,000,000 inhabitants, and her settlements two and a half centuries old, has 748,000 oxen and stock cattle; while Pennsylvania, with more than 3,000,000 people, has 721,000 cattle; while Ohio, with 3,000,000 people, has 749,000 cattle; while Illinois, with 2,800,000 people, has 867,000 cattle; and while Iowa, with 1,200,000 people, has 686,000 cattle; Texas, forty years of age, and with her 500,000 people, had 2,000,000 head of oxen and other cattle, exclusive of cows, in 1867, as shown by the returns of the county assessors.

" In 1870, allowing for the difference between the actual number of cattle owned and the number returned for taxation, there must be fully 3,000,000 head of beeves and stock cattle. This is exclusive of cows, which, at the same time, are reported at 600,000 head. In 1870 they must number 800,000—making a grand total of 3,800,000 head of cattle in Texas. One-fourth of these are beeves, one-fourth are cows, and the other two-fourths are yearlings and two-year olds. There would, therefore, be 950,000 beeves, 950,000 cows, and 1,900,000 young cattle. There are annually raised and branded 750,000 calves. These cattle are raised on the great plains of Texas, which contain 152,000,000 acres. In the vast regions watered by the Rio Grande, Nueces, Guadalupe, San Antonio, Colorado, Leon, Brazos, Trinity, Sabine, and Red Rivers, these millions of cattle graze upon almost tropical growths of vegetation. They are owned by the ranchmen, who own from 1,000 to 75,000 head each."

As specimen ranches, may be named the following: Santa Catrutos Ranch belongs to Richard King. Amount of land, 84,132 acres. The stock consists of 65,000 cattle, 10,000 horses, 7,000 sheep, 8,000 goats. Three hundred Mexicans are employed, and 1,000 saddle horses, on the place. O'Connor's ranch, near Goliad, is an estate possessing about 50,000 cattle. The Robideaux ranch, on the Gulf, belonging to Mr. Kennedy, contains 142,840 acres of land, and has 30,000 beef cattle in addition to other stock.

THE CLIMATE OF THE PLAINS.

Mr. R. S. Elliott, who has studied this matter carefully, says: "The plains have been so often described as a rainless region that great misconception in regard to the climate has prevailed. The absolute precipitation is much greater than has been in past years supposed, and is due to other causes. Meteorologists who have described the rain-fall of the plains as derived only or principally from the remaining moisture of winds from the Pacific, after the passage of the Nevada and Rocky Mountain ranges, have been greatly in error, and the better conclusion now is, with all authorities who have given any special attention to the subject, that the moisture which fertilizes the Mississippi Valley, including the broad, grassy plains, is derived from the Gulf of Mexico.

"At Fort Riley about sixty-nine per cent. of the annual precipitation is in spring and summer; at Fort Kearney, eighty-one; and at Fort Laramie, seventy-two per cent. From observations at Forts Harker, Hays, and Wallace, on the line of this road, the same rule seems to hold good. Records have not been long enough continued at these three posts to give a long average, but the mean appears to be between seventeen and nineteen inches at Hays and Wallace, and possibly rather more at Harker. The actual average for 1868 and 1869 at Hays is 18.76 inches, and for the first six months of 1870 the record is 10.68 inches. At Wallace the record for 1869 was over seventeen inches, and in 1870, up to October 1, about the same amount had fallen.

"Without records there can be only conjecture; and I can only remark that there does not seem to be much diminution in the annual rain-fall until we get as far west as the one hundred and third meridian. Thence to the base of the mountains (except perhaps in the timbered portions of the great divide south of the line of this railway) the annual average may be possibly two or three inches less than in the midst of the plains—a peculiarity explained, hypothetically, by the fact that the region 'lies to the westward of the general course of the moisture currents of air flowing northward from the Gulf of Mexico, and is so near the mountains as to lose much of the precipitation that localities in the plains east and north-east are favored with. The mountains seem to exercise an influence—electrical and magnetical—in attracting moisture, which is condensed in the cooler regions of their summits, while the plains at their feet may be parched and heated to excess.' This explanation may be fanciful, but the fact remains that near the mountains the rains seem to decrease north of the great divide; fortunately, however, this occurs in a region where irrigation may be applied extensively and where there is sufficient moisture to nourish bountiful crops of grass.

"The vegetation of the plains along wagon tracks and rail road embankments shows a capability of production scarcely suggested by the surface where undisturbed: wherever the earth is broken up, the wild sunflower (*Helianthus*), and others of the taller-growing plants, though previously unknown in the vicinity, at once spring up.

"I have been on the plains all the time since early in May till this date (22d of September). There has been much dry weather, but I have not seen one cloudless day—no day on which the sun would rise clear and roll along a canopy of brass to the west. There has always been humidity enough to form clouds at the proper height; and on many days they would be seen defining, by their flat bottoms, the exact line where condensation became sufficient to render the vapor visible. I conclude, from all this, that abundant moisture has floated over the plains to have given us a great deal more rain than would be desirable if it had been precipitated.

"Sometimes a storm would be seen to gather near the horizon, and we could see the rain pending from the clouds like a fringe, hanging apparently in mid-air, unable to reach the expectant earth. The rain stage of condensation had been reached above, but the descending shower was re-vaporized apparently, and thus arrested.

"These hot winds are not, so far as I have observed, apt to be constant in one place for any considerable length of time; they strike your face suddenly, and perhaps in a minute are gone. They seem to run along in streaks or *ovenfulls* with the winds of ordinary (but rather high) temperature. They do not begin, I believe, till in July, as a general rule, and are over by September 1, or perhaps by August 15. Their origin I take to be, of course, in heated regions south or southwest of us; but their peculiar occurrence, so capricious and often so brief, I can not explain to myself satisfactorily.

"I may remark that this season, since about the 15th of July, in these distant plains, has given us rain enough to make beautifully verdant the spots in the prairie burnt off during the "heated" term in July. From Kit Carson eastward, the rains have been, I think, exceptionally abundant. All through the summer we have had *dew* occasionally, and it has been remarked that buffalo meat has been more difficult of preservation than heretofore—facts indicative of humidity in the atmosphere, even where but little rain-fall was witnessed. Turnips sown in August would have made a crop in this vicinity—four hundred and twenty-two miles west of the state line of Missouri."

CLIMATIC CHANGES ON THE PLAINS.

"Facts such as these," continues the same writer, "seem to sustain the popular persuasion that a *climatic change* is taking place, promoted by the spread of settlements westwardly, breaking up portions of the prairie soil, covering the earth with plants that shade the ground more than the short grasses; thus checking or modifying the reflection of heat from the earth's surface, etc. The fact is also noted that even where the prairie soil is not disturbed, the short buffalo grass disappears as the "frontier" extends westward, and its place is taken by grasses and other herbage of taller growth. That this change of the clothing of the plains, if sufficiently extensive, might have a modifying influence on the climate, I do not doubt; but whether the change has been already spread over a large enough area, and whether our apparently or really wetter seasons may not be part of a cycle, are unsettled questions.

"The civil engineers of the railways believe that the rains and humidity of the plains have increased during the extension of railroads and telegraphs across them. If this is the case, it may be that the mysterious electrical influence in which they seem to have faith, but do not profess to explain, has exercised a beneficial influence. What effect, if any, the digging and grading, the iron rails, the tension of steam in locomotives, the friction of metallic surfaces, the poles and wires, the action of batteries, etc., could possibly or probably have on the electrical conditions, as connected with the phenomena of precipitation, I do not, of course, undertake to say. It may be that wet seasons have merely happened to coincide with railroads and telegraphs. It is to be observed that the poles of the telegraph are quite frequently destroyed by lightning; and it is probable that the lightning thus strikes in many places where before the erection of the telegraph it was not apt to strike, and perhaps would not reach the earth at all.

"It is certain that rains have increased; this increase has coincided with the extension of settlements, railroads, and telegraphs. If influenced by these, the change of climate will go on; if by extra mundane influences, the change may be permanent, progressive, or retrograde. I think there are good grounds to believe it will be progressive. Within the last fifteen years, in Western Missouri and Iowa, and in Eastern

Kansas and Nebraska, a very large aggregate surface has been broken up, and holds more of the rains than formerly. During the same period modifying influences have been put in motion in Montana, Utah, and Colorado. Very small areas of timbered land west of the Missouri have been cleared—not equal, perhaps, to the area of forest, orchard, and vineyards planted. Hence it may be said that all the acts of man in this vast region have tended to produce conditions on the earth's surface ameliorative of the climate. With extended settlements on the Arkansas, Canadian, and Red River of the south, as well as on the Arkansas, on the river system of the Kaw Valley, and on the Platte, the ameliorating conditions will be extended in like degree; and it partakes more of sober reason than wild fancy to suppose that a permanent and beneficial change of climate may be experienced. The appalling deterioration of large portions of the earth's surface, through the acts of man in destroying the forests, justifies the trust that the culture of taller herbage and trees in a region heretofore covered mainly by short grasses may have a converse effect. Indeed, in Central Kansas nature seems to almost precede settlements by the taller grasses and herbage."

THE TREES AND FUTURE FORESTS OF THE PLAINS.

Mr. Elliott continues his article as follows: "The principal native trees on the plains west of ninety-seventh meridian are: Cottonwood, walnut, elm, ash, box-elder, hackberry, plum, red cedar. To these may be added willow and grape-vines, and also the locust and wild cherry mentioned by Abert as occurring on the Purgatory. The black walnut extends to the one-hundredth meridian. The elm and ash are of similar, perhaps greater range. Hackberry has been observed west of one hundred and first meridian. Cottonwood, elder, red cedar, plum, and willow are persistent to the base of the mountains. The extensive pine forest on the 'great divide' south of Denver, although stretching seventy to eighty miles east from the mountains, is not taken into view as belonging to the plains proper. Its existence, however, suggests the use of its seeds in artificial plantations in that region. The fossil wood imbedded in the cretaceous strata in many parts of the plains is left out of consideration, as belonging to a previous, though recent, geological age; but the single specimens of trees found growing at wide intervals are silent witnesses to the *possibility* of extended forest growth.

"Were it possible to break up the surface to a depth of two feet, from the ninty-seventh meridian to the mountains, and from the thirty-fifth to the forty-fifth parallel, we should have in a single season a growth of taller herbage over the entire area, less reflection of the sun's heat, more humidity in the atmosphere, more constancy in springs, pools, and streams, more frequent showers, fewer violent storms, and less caprice and fury in the winds. A single year would witness a changed vegetation and a new climate. In three years (fires kept out) there would be young trees in numerous places, and in twenty years there would be fair young forests. The description of the 'broad, grassy plains,' given in the foregoing pages, attests their capacity to sustain animal life. For cattle, sheep, horses, and mules, they are a natural pasture in summer, with (in many parts) hay cured standing for winter. The famed Pampas, with their great extremes of wet and drought, can not bear comparison with our western plains. For grazing purposes, the habitable character of our vast traditional 'desert' is generally conceded, and hence it need not be enlarged on here"

THE SUPPLY OF FUEL.

Of the question of fuel for the future dwellers upon the face of Buffalo Land, Hayden, in his report, speaks as follows:

"The question often arises in the minds of visitors to this region, how the law of compensation supplies the want of fuel in the absence of trees for that use. Many persons have taken the position that the Creator never made such a vast country, with a soil of such wonderful fertility, and rendered it so suitable for the abode of man, without storing in the earth beds of carbon for his needs. If this idea could be shown to be true in any case, we would ask why are the immense beds of coal stored away in the mountains of Pennyslvania and Virginia, while at the same time the surface is covered with dense forests of timber. We now know that this law does not apply to the natural world; and, if it did, this western country would be a remarkable exception. The State of Nebraska seems to be located on the western rim of the great coal basin of the West, and only thin seams of poor coal will probably ever be found; but in the vicinity of the Rocky Mountains, in Wyoming, and Colorado, coal in immense quantities has been hidden away for ages, and the Union Pacific Railroad has now brought it near the door of every man's dwelling.

"These Rocky Mountain coal-beds will one day supply an abundance of fuel for more than one hundred thousand square miles along the Missouri River of the most fertile agricultural land in the world."

Of this coal area, Persifor Frazier, Jr., says: "Those beds which occur on the east flank of the Rocky Mountains have been followed for five hundred miles and more, north and south; and if it be true that these are 'fragments of one great basin, interrupted here and there by the upheaval of mountain chains, or concealed by the deposition of newer formations,' then their extension east and west, or from the eastern range of the Rocky Mountains or Black Hills to Weber Canyon, where an excellent coal is mined, will fall but little short of five hundred miles. Throughout this extent these beds of coal are found between the upper cretaceous and lower tertiary (or in the transition beds of Hayden), wherever these transition beds occur, whether on the extreme flanks or in the valleys and parks between the numerous mountain ranges. Assuming that the eroding agencies together have cut off one-half of the coal from this area, and taking one-half of the remainder as their average longitudinal extent, we have over fifty thousand square miles of coal lands, accounting the latitudinal extent as only five hundred miles; whereas we have no reason to believe that it terminates within these bounds, but, on the contrary, good reason for supposing that it extends northward far into Canada, and southward with the Cordilleras. All this territory has been omitted in the estimate of the extent of our coal fields."

DISTRICTS CONTIGUOUS TO THE PLAINS.

The reader has now had the salient features of the great plains placed before him in succession. The more interesting districts immediately adjoining will well repay the reader for a brief consideration.

THE NORTH PLATTE DISTRICT.

A late writer, who has studied the country of which he speaks very closely,* thus describes the North Platte District:

"The distance from the mouth of the North Platte, where it joins the South Platte on the Union Pacific Railroad, to its sources in the great Sierra Madre, whose lofty sides form the North Park, in which this stream takes its rise, is more than eight hundred miles. Its extreme southern tributaries head in the gorges of the mountains one hundred miles south of the railroad, and receive their water from the melting snows of these snow-capped ranges. Its extreme western tributaries rise in the Wahsatch and Wind River ranges, sharing the honor of conveying the crystal snow waters from the continental divide with the Columbia and Colorado of the Pacific. Its northern tributaries start oceanward from the Big Horn Mountains, three hundred miles north of the starting-point of its southern sources.

"It drains a country larger than all New England and New York together. East of the Alleghany Mountains there is no river comparable to this clear, swift mountain stream in its length or in the extent of country it drains.

"The main valley of the North Platte, two hundred miles from its mouth to where it debouches through the Black Hills out on to the great plains, is an average of ten miles wide. Nearly all this area—two thousand square miles—is covered

*Dr. H. Latham, under date June 5th, 1870, in the Omaha Daily Herald.

with a dense growth of grass, yielding thousands of tons of hay. The bluffs bordering these intervals are rounded and grass-grown, gradually smoothing out into great grassy plains, extending north and south as far as the eye can see.

"Of the country, Alexander Majors says, in a letter to the writer of this article: 'The favorite wintering ground of my herders for the past twenty years has been from the Caché a la Poudre on the south to Fort Fetterman on the north, embracing all the country along the eastern base of the Black Hills.' It was of this country that Mr. Seth E. Ward spoke, when he says: 'I am satisfied that no country in the same latitude, or even far south of it, is comparable to it as a grazing and stock-raising country. Cattle and stock generally are healthy, and require no feeding the year round, the rich 'bunch' and 'gramma' grasses of the plains and mountains keeping them, ordinarily, fat enough for beef during the entire winter.'

"All this region east of the Black Hills is at an elevation less than five thousand feet. The climate, as reported from Fort Laramie for a period of twenty years, is 50° Fahrenheit. The mean temperature for the spring months is 47°, for the summer months 72°, for autumn 60°, for winter 31°. The annual rain-fall is about eighteen inches—distributed as follows: Spring, 8.69 inches; summer, 5.70 inches; autumn, 3.69 inches. The snow fall is eighteen inches.

"There is in the North Platte Basin, east of the Black Hills divide, at least eight million acres of pasturage, with the finest and most lasting streams, and good shelter in the bluffs and canyons. As I have said before, we can only judge of the extent and resources of such a single region by comparison. Ohio has six million sheep, yielding eighteen million pounds of wool, bringing herd farmers an aggregate of four and one-half million dollars. This eight million acres of pasture would at least feed eight million sheep, yielding twenty-four million pounds of wool, and, at the same price as Ohio wool, six million dollars. Now, this money, instead of going to build up ranches, stock-farms, store-houses, woolen mills, and all the components of a great and thrifty settlement, is sent by our wool-growers and woollen manufacturers to Buenos Ayres, to Africa, and Australia, to enrich other people and other lands, while our wool-growing resources remain undeveloped.

"As you follow the North Platte up through the Black Hill Canyon, you come out upon the great Laramie plains, which lie between the Black Hills on the east and the snowy range on the west. These plains are ninety miles north and south, and sixty miles east and west. They are watered by the Big and Little Laramie Rivers, Deer Creek, Rock Creek, Medicine Bow River, Cooper Creek, and other tributaries of the North Platte. It is on the extreme northern portion of these plains, in the valley of Deer Creek, that General Reynolds wintered during the winter of 1860, and of which he remarks, on pages seventy-four and seventy-five of his 'Explorations of the Yellowstone," as follows:

"'Throughout the whole season's march the subsistence of our animals had been obtained by grazing after we had reached our camp in the afternoon, and for an hour or two between the dawn of day and our time of starting. The consequence was that when we reached our winter quarters there were but few animals in the train that were in a condition to have continued the march without a generous grain diet. Poorer and more broken-down creatures it would be difficult to find. In the spring they were in as fine condition for commencing another season's work as could be desired. A greater change in their appearance could not have been produced even if they had been grain-fed and stable-housed all winter. Only one was lost, the furious storm of December coming on before it had gained sufficient strength to endure it. The fact that seventy exhausted animals, turned out to winter on the plains the first of November, came out in the spring in the best condition, and with the loss of but one of their number, is the most forcible commentary I can make on the quality of the grass and the character of the winter.'

"These plains have been favorite herding grounds of the buffalo away back in the pre-historic age of this country. Their bones lie bleaching in all directions, and their paths, deeply worn, cover the whole plain like a net-work. Their 'wallows,' where these shaggy lords of animal creation tore deep pits into the surface of the ground, are still to be seen. Elk, antelope, and deer still feed here, and the mountain sheep are found on the mountain sides and in the more secluded valleys of the Sierra Madre range—all proving conclusively that this has afforded winter pasturage from time immemorial. Since 1849 many herds of work-oxen, belonging to emigrants, freighters, and ranchmen, have grazed here each winter.

"South of the Laramie plains is the North Park, one of three great parks of the Rocky Mountains, so fully described by Richardson, Bross, and Bowles. This North Park is formed by the great Snowy Range. It is a valley from six to eight thousand feet high, ninety miles long, and forty miles wide, surrounded by snowy mountains from thirteen to fifteen thousand feet high. These mountain tops and sides are completely covered with dense growths of forests; the lower hillsides and this great valley are covered with grasses. The forests and mountains afford ample shelter from sweeping winds. Here, as well as on the Laramie plains, the buffalo grazed in great herds; and here the Ute hunters, from some hidden canyons, dashed down among them on their trained and fleet ponies, shooting their arrows with unerring aim on all sides, and having such glorious sport as kings might court and envy. The Indians are now gone from this valley, and the buffalo nearly so. On the two million acres in this valley not twenty head of cattle graze.

"This great park, splendidly watered by the three forks of the Platte, and by a hundred small streams that drain these lofty mountains of their snows and rains—rich in all kinds of nutritious grasses, plentifully supplied with timber; on the tertiary coal fields, with iron, copper, lead, and gold—has not one real settler. There are a few miners, but where there should be flocks and herds of sheep and cattle without number, there is only the wild game—the elk, antelope, and deer."

THE VALLEYS OF THE WHITE EARTH AND NIOBRARA.

These streams are branches of the Missouri—the one mainly in Dakota Territory, and the other in Nebraska. The following graphic paragraphs concerning them are from Hayden again:

"I have spent many days exploring this region (the White Earth Valley) when the thermometer was 112° in the shade, and there was no water suitable for drinking purposes within fifteen miles. But it is only to the geologist that this place can have any permanent attraction. He can wind his way through the wonderful canyons among some of the grandest ruins in the world. Indeed, it resembles a gigantic city fallen to decay. Domes, towers, minarets and spires may be seen on every side, which assume a great variety of shapes when viewed in the distance. Not unfrequently the rising or the setting sun will light up these grand old ruins with a wild, strange beauty, reminding one of a city illuminated in the night, when seen from some high point. The harder layers project from the sides of the valley or canyon with such regularity that they appear like seats, one above the other, of some vast amphitheater.

"It is at the foot of these apparent architectural remains that the curious fossil treasures are found. In the oldest beds we find the teeth and jaws of a Hyopotamus, a river-horse much like the hippopotamus, which must have sported in his pride in the marshes that bordered this lake. So, too, the Titanotherum, a gigantic pachyderm, was associated with a species of hornless rhinoceros. These huge rhinoceroid animals appear at first to have monopolized this entire region, and the plastic, sticky clay of the lowest bed of this basin, in which the remains were found, seems to have formed a suitable bottom of the lake in which these thick-skinned monsters could wallow at pleasure."

Of the *fauna* of the Niobrara and Loup Fork Valleys, he speaks as follows: "In the later fauna were the remains of a number of species of extinct camels, one of which was of the size of the Arabian camel, a second about two-thirds as large. Not less interesting are the remains of a great variety of forms of the horse family, one of which was about as large as the ordinary domestic animal, and the smallest not more than two or two and a half feet in height, with every intermediate grade in size."

NEW MEXICO—ITS SOIL, CLIMATE, RESOURCES, ETC.

Bordering on what might be called the south-western corner of the plains, or perhaps more properly forming, over its eastern half, part of them, lies New Mexico. I find the following valuable description of the soil, climate, and productions of this section in the report of Prof. Cyrus Thomas:

"The best estimate I can make of the arable area of the Territory is about as follows: In the Rio Grande district, one twentieth, or about two thousand eight hundred square miles; in the strip along the western border, one-fiftieth, or about six hundred square miles; in the north-eastern triangle, watered by the Canadian River, one-fifteenth, or about one thousand four hundred square miles. This calculation excludes the 'Staked Plains,' and amounts in the aggregate to four thousand eight hundred square miles, or nearly two million nine hundred thousand acres. This, I am aware, is larger than any previous estimate that I have seen; but when the country is penetrated by one or two railroads, and a more enterprising agricultural population is introduced, the fact will soon be developed that many portions now considered beyond the reach of irrigation will be reclaimed. I do not found this estimate wholly upon the observations made in the small portions I have visited, but, in addition thereto, I have carefully examined the various reports made upon special sections, and have obtained all the information I could from intelligent persons who have resided in the Territory for a number of years.

"As the Territory includes in its bounds some portions of the Rocky Mountain range on which snow remains for a great part of the year, and also a semi-tropical region along its southern boundary, there is, of necessity, a wide difference in the extremes of temperature. But, with the exception of the cold seasons of the higher lands at the north, it is temperate and regular. The summer days in the lower valleys are quite warm, but, as the dry atmosphere rapidly absorbs the perspiration of the body, it prevents the debilitating effect experienced where the air is heavier and more saturated with moisture. The nights are cool and refreshing. The winters, except in the mountainous portions at the north, are moderate, but the difference between the northern and southern sections during this season is greater than during the summer. The amount of snow that falls is light, and seldom remains on the ground longer than a few hours. The rains principally fall during the months of July, August, and September, but the annual amount is small, seldom exceeding a few inches. When there are heavy snows in the mountains during the winter, there will be good crops the following summer, the supply of water being more abundant, and the quantity of sediment carried down greater, than when the snows are light. Good crops appear to come in cycles—three or four following in succession; then one or two inferior ones.

"During the autumn months the wind is disagreeable in some places, especially near the openings between high ridges, and at the termini of or passes through mountain ranges. There is, perhaps, no healthier section of country to be found in the United States than that embraced in the boundaries of Colorado and New Mexico; in fact, I think I am justified in saying that this area includes the healthiest portion of the Union. Perhaps it is not improper for me to say that I have no personal ends to serve in making this statement, not having one dollar invested in either of these Territories in any way whatever; I make it simply because I believe it to be true.

Nor would I wish to be understood as contrasting with other sections of the Rocky Mountain region, only so far as these Territories have the advantage in temperature. It is possible Arizona should be included, but, as I have not visited it, I can not speak of it.

"There is no better place of resort for those suffering with pulmonary complaints than here. It is time for the health-seekers of our country to learn and appreciate the fact that within our own bounds are to be found all the elements of health that can possibly be obtained by a tour to the eastern continent, or any other part of the world; and that, in addition to the invigorating air, is scenery as wild, grand, and varied as any found amid the Alpine heights of Switzerland. And here, too, from Middle Park to Los Vegas, is a succession of mineral and hot springs of almost every character.

"The productions of New Mexico, as might be inferred from the variety of its climate, are varied, but the staples will evidently be cattle, sheep, wool, and wine, for which it seems to be peculiarly adapted. The table-lands and mountain valleys are covered throughout with the nutritious gramma and other grasses, which, on account of the dryness of the soil, cure upon the ground, and afford an inexhaustible supply of food for flocks and herds both summer and winter. The ease and comparatively small cost with which they can be kept, the rapidity with which they increase, and exemption from epidemic diseases, added to the fact that winter-feeding is not required, must make the raising of stock and wool-growing a prominent business of the country—the only serious drawback at present being the fear of the hostile Indian tribes. But, as these remarks apply equally well to all these districts, I will speak further in regard to this matter when I take up the subject of grazing in this division.

"The cattle and sheep of this Territory are small, because no care seems to be taken to improve the breed. San Miguel County appears to be the great pasturing ground for sheep, large numbers being driven here from other counties to graze. Don Romaldo Baca estimates that between five hundred thousand and eight hundred thousand are annually pastured here—about two-thirds of which are driven in from other sections. His own flocks number between thirty thousand and forty thousand head; those of his nephew twenty-five thousand to thirty thousand; Mr. Mariano Trissarry, of Bernalillo County, owns about fifty-five thousand; and Mr. Gallegos, of Santa Fé, nearly seventy thousand head.

"Don Romaldo Baca stated to me that his flocks yielded him an annual average of about one and a half pounds of washed wool to the sheep; that the average price of sheep was not more than two dollars per head; that the wool paid all expenses, and left the increase, which is from fifty to seventy-five per cent. per annum, as his profit. From these figures some estimate may be formed of what improved sheep would yield.

"Wheat and oats grow throughout the Territory, but the former does not yield as heavily in the southern as in the northern part. If any method of watering the higher plateau is ever discovered, I think that it will produce heavier crops of wheat than the Valley of the Rio Grande.

"Corn is raised from the Vermijo, on the east of the mountains, around to the Culebra, on the inside; in fact, it is the principal crop of San Miguel County, but the quality and yield is inferior to that which can be produced in the Rio Grande Valley and along the Rio Bonito. The southern portion of the Rio Pecos Valley and the Canadian bottoms are probably the best portions of the Territory for this cereal.

"Apples will grow from the Taos Valley south, but peaches can not be raised to any advantage north of Bernalillo, in the central section; but it is likely they would do well along some of the tributaries and main valley of the Canadian River. They also appear to grow well and produce fruit without irrigation in the Zuñi country; and the valley of the Mimbres is also adapted to their culture. Apricots and plums grow wherever apples or peaches can be raised. I neglected to obtain any information in regard to pears, but, judging from the similarity of soil and climate here to that of Utah and California, where this fruit grows to perfection, I suppose that in the central and southern portions it would do well.

"The grape will probably be the chief, or at least the most profitable, product of the soil. The soil and climate appear to be peculiarly adapted for its growth, and the probability is that, as a grape-growing and wine-producing section, it will be second only to California. From Col. McClure I learned that the amount of wine made in 1867 was about forty thousand gallons, and that the crop of 1869 would probably reach one hundred thousand gallons. I have not been informed since whether his estimate was verified or not. A good many vineyards were planted in 1869—at least double the number of 1868. Several Americans, anticipating the building of a railroad through that section, have engaged in this branch of agriculture. The wine that is made here is said to be of an excellent quality.

"Beets here, as in Colorado, grow to an enormous size, and it is quite likely that the sugar beet would not only yield heavy crops, but also contain a large per cent. of saccharine matter. I am rather inclined to believe that soil which is impregnated with alkaline matter will favor the production of the saccharine principle. I base this opinion wholly on observations made in Utah in regard to its effect on fruit; therefore experiments may prove that I am wholly mistaken. It is possible the experiment has been tried; if so, I am not aware of it.

"The Irish potatoes are inferior to those raised further north. Cabbages grow large and fine. Onions from the Raton Mountains south have the finest flavor of any I ever tasted, and therefore I am not surprised that Lieut. Emory found the dishes at Bernalillo 'all dressed with the everlasting onion.' But, as to the 'Chili,' or pepper, which is so extensively raised and used in New Mexico, I beg to be excused, unless I can have my throat lined with something less sensitive than nature's coating. Sweet potatoes have been successfully tried in the vicinity of Fort Sumner and along the head-waters of the Rio Bonito. Melons, pumpkins, frijoles, etc., are raised in profusion in the lower valleys; and I understand cotton was formerly grown in limited quantities.

"As a general thing, the mountains afford an abundance of pine for the supply of lumber and fuel to those sufficiently near to them. Some of the valleys have a limited amount of cottonwood growing along them. In addition to pine, spruce and cottonwood, the stunted cedar and mesquit, which is found over a large area, may be used for fuel. The best timbered portion of the Rio Grande Valley is between Socorro and Doña Aña. The east side of the Guadalupe range has an abundant supply of pine of large size. Around the head-waters of the Pecos is some excellent timber. Walnut and oak are found in a few spots south, but in limited quantities, and of too small a size to be of much value."

THE DISAPPEARING BISON.

In connection with this general review of Buffalo Land, it is interesting to note that while civilization, advancing from the east, pushes our bison west, another tide of human beings, creeping out from the mountains eastward, presses the buffalo back before it. The brute multitude is thus between two advancing lines, which will soon crush it. In confirmation of this, I find the following in Hayden's notes of the country along the base of the Laramie Mountains:

"These broad, grassy plains are not yet entirely destitute of their former inhabitants; flocks of antelope still feed on the rich, nutritious grasses; but the buffalo, which once roamed here by thousands, have disappeared forever. No trace of them is now left but the old trails, which pass across the country in every direction, and the bleached skulls which are scattered here and there over the ground. These traces are fast passing away. The skulls are decaying rapidly, and this once peculiar feature of the landscape in the West will be lost. Two years ago I collected a large quantity of these bleached skulls and distributed them to several of our museums, in order to insure their preservation.

"There is also a singular ethnological fact connected with these skulls. We shall observe that the greater part of them have the forehead broken in for a space of three or four inches in diameter. Whenever an Indian kills a buffalo, he fractures the skull with his tomahawk and extracts the brains, which he devours in a raw state.

"Indians or old trappers traveling through the enemy's country always fear to build a fire, lest the smoke attract the notice of the foe. The consequence is that they have contracted the habit of eating certain parts of an animal in an uncooked condition. I have estimated that six men may make a full meal from a buffalo without lighting a fire. The ribs on one side are taken out with a knife, and the concavity serves as a dish. The brains are taken out of the skull, and the marrow from the leg-bones, and the two are chopped together in the rib-dish. The liver and lungs are eaten with a keen relish; also certain portions of the intestines; and the blood supplies an excellent and nutritious drink.

"Both Indian and buffalo have probably disappeared forever from these plains. Elk, black-tailed deer, red deer, mountain sheep, wolves, and the smaller animals, are still quite abundant, especially in the valleys of the small streams, where they flow down through the mountains. Elk Mountain and Sheephead Mountain have always been noted localities for these animals."

THE FISH WITH LEGS.

But while the buffalo has become extinct in that locality, an inhabitant of the water may be preparing (query: in support of the theory of development?) to take its place. I quote again from Hayden:

"There are other attractions here, of which the traveler will be informed long before he reaches the locality. The 'fish with legs' are the only inhabitants of the lake, and numbers of persons make it a business to catch and sell them to travelers. During the summer season they congregate in great numbers in the shallow water among the weeds and grass near the shore, and can be easily caught; but in cold weather they retire to the deeper portions of the lake, and are not seen again until spring. These little animals are possessed of gills, and, were it not for the legs, would most nearly resemble a miniature cat-fish. But when warm weather comes, a form closely resembling them, but entirely destitute of gills, may be seen in the water swimming, or creeping clumsily about on land. Sometimes they travel long distances, and are found in towns, near springs or wet places, usually one at a time, while those with gills are never seen except in the alkaline lakes which are so common all over the West."

THE MOUNTAIN SUPPLY OF LUMBER FOR THE PLAINS.

In connection with this (the western) border of the plains, it is interesting to note what the same writer says of a future supply of lumber:

"Not only in the more lofty ranges, but also in the lower mountains, are large forests of pine timber, which will eventually become of great value to this country. Vast quantities of this pine, in the form of railroad ties, are floated down the various streams to the Union Pacific Railroad. One gentleman alone contracted for five hundred and fifty thousand ties, all of which he floated down the stream from the mountains along the southern side of the Laramie Plains. The Big and Little Laramie, Rock Creek, and Medicine Bow River, with their branches, were here literally filled with ties at one time; and I was informed that, in the season of high water, they can be taken to the railroad from the mountains, after being cut and placed in the water, at the rate of from one to three cents each. These are important facts, inasmuch as they show the ease with which these vast bodies of timber may be brought to the plains below and converted into lumber, should future settlement of the country demand it."

* * * * * * * *

"On the summits of these lofty mountains are some most beautiful, open spots, without a tree, and covered with grass and flowers. After passing through dense pine forests for nearly ten miles, we suddenly emerged into one of these park-like areas. Just in the edge of the forest which skirted it were banks of snow six feet deep, compact like a glacier, and within a few feet were multitudes of flowers—and even the common strawberry seemed to flourish. These mountains are full of little streams of the purest water, and for six months of the year good pasturage for stock could be found."

THE END.

www.ingramcontent.com/pod-product-compliance
Lightning Source LLC
LaVergne TN
LVHW021320110826
845150LV00003B/624

9781425556327